CELLULAR AUTOMATA

BICENTENNIAL

1807 WILEY 2007

BICENTENNIAL

THE WILEY BICENTENNIAL–KNOWLEDGE FOR GENERATIONS

Each generation has its unique needs and aspirations. When Charles Wiley first opened his small printing shop in lower Manhattan in 1807, it was a generation of boundless potential searching for an identity. And we were there, helping to define a new American literary tradition. Over half a century later, in the midst of the Second Industrial Revolution, it was a generation focused on building the future. Once again, we were there, supplying the critical scientific, technical, and engineering knowledge that helped frame the world. Throughout the 20th Century, and into the new millennium, nations began to reach out beyond their own borders and a new international community was born. Wiley was there, expanding its operations around the world to enable a global exchange of ideas, opinions, and know-how.

For 200 years, Wiley has been an integral part of each generation's journey, enabling the flow of information and understanding necessary to meet their needs and fulfill their aspirations. Today, bold new technologies are changing the way we live and learn. Wiley will be there, providing you the must-have knowledge you need to imagine new worlds, new possibilities, and new opportunities.

Generations come and go, but you can always count on Wiley to provide you the knowledge you need, when and where you need it!

William J. Pesce
WILLIAM J. PESCE
PRESIDENT AND CHIEF EXECUTIVE OFFICER

Peter Booth Wiley
PETER BOOTH WILEY
CHAIRMAN OF THE BOARD

CELLULAR AUTOMATA

A Discrete View of the World

Joel L. Schiff

University of Auckland

BICENTENNIAL 1807 WILEY 2007 BICENTENNIAL

WILEY-INTERSCIENCE

A JOHN WILEY & SONS, INC., PUBLICATION

Published by John Wiley & Sons, Inc., Hoboken, New Jersey.
Published simultaneously in Canada.

Wiley Bicentennical Logo: Richard J. Pacifico

Library of Congress Cataloging in Publication Data:

Schiff, Joel L.
Cellular automata : a discrete view of the world / Joel L. Schiff
p. cm. -- (Wiley series in discrete mathematics and optimization)
Includes bibliographical references and index.

ISBN 978-0-470-16879-0 (cloth)

Printed in the United States of America

10 9 8 7 6 5 4 3 2 1

Dedicated to my beloved wife, Christine

Life is a peephole, a single tiny entry onto a vastness — how can I not dwell on this brief, cramped view I have of things?

— *From* Life of Pi *by Yann Martel*

CONTENTS

PREFACE

... synthetic universes defined by simple rules...

— *Tommaso Toffoli and Norman Margolus* – Cellular Automata Machines

I expect the children of 50 years from now will learn cellular automata before they learn algebra.

— *Stephen Wolfram* – New Scientist, *November 18, 2006*

The history of cellular automata is only quite recent, coming to life at the hands of two fathers, John von Neumann and Stanislaw Ulam in the early 1950s. Subsequent work in the early 1960s included that of Ulam and his co-workers at Los Alamos and by John Holland at the University of Michigan whose work on adaptation continued for several decades. Early theoretical research was conducted by Hedlund, Moore, and Myhill, among many others, not always under the name of cellular automata (CA), since the concept was still in its formative stages. A big boost to the popularization of the subject came from John Conway's highly addictive Game of Life presented in Martin Gardner's October 1970 column in *Scientific American*. Still the study of CA lacked much depth, analysis, and applicability and could not really be called a scientific discipline.

All that changed in the early 1980s when physicist Stephen Wolfram in a seminal paper, "Statistical mechanics of cellular automata," initiated the first serious study of cellular automata. In this work and in a series of subsequent ones Wolfram began producing some of the images that have now become iconic in the field. Conferences were organized and people from various disciplines were being drawn into the field. It is now very much an established scientific discipline with applications found in a great many areas of science. Wolfram has counted more than 10,000 papers referencing his original works on the subject and the field of cellular automata has taken on a life of its own. Another milestone in the field was the publication in 2002 of *A New Kind of Science*, Wolfram's 1200 page *magnum opus*, a monumental work which brought the subject to the attention of a truly global audience.

The CA paradigm is very appealing and its inherent simplicity belies its potential complexity. Simple local rules govern an array of cells that update the state they are in at each tick of a clock. It has been found that this is an excellent way to analyze a great many natural phenomena, the reason being that most physical processes are themselves local in nature — molecules interact locally with their neighbors, bacteria with their neighbors, ants with theirs, and people likewise. Although natural phenomena are also continuous, examining the system at discrete time steps does not really diminish the power of the analysis. So in the artificial CA world we have an unfolding microcosm of the real world.

One of the things self-evident to everyone is the order that is found in Nature. From an ameoba to plants to animals to the universe itself, we find incredible order everywhere. This begs the obvious questions: Where did this order come from — how could it have originated? One of the fundamental lessons of cellular automata is that they are capable of self-organization. From simple local rules that say nothing whatsoever about global behavior, we find that global order is nonetheless preordained and manifest in so many of the biological and physical systems that we will consider. In the words of theoretical biologist Stuart Kauffman, it is "order for free." It is this order for free that allows us to emulate the order we find in Nature.

Related to the creation of order is the notion of complexity. How can a finite collection of chemicals make up a sentient human being? Clearly the whole is greater than the sum of its parts. How can termites build complex structures when no individual termite who starts a nest even lives to see its completion? The whole field of complexity has exploded over recent years and here too CA play their part. One of the most endearing creatures that we shall encounter is Langton's ant in Chapter 6, and this little creature will teach us a lot about complexity.

Of course it is no longer possible in a single text to cover every aspect of the subject. The field, as Wolfram's manuscript count shows, has simply grown too large. So this monograph is merely an introduction into the brave new world of cellular automata, hitting the highlights as the author sees them. The mathematics in Chapter 1 serves mainly to relate the notions of fractals, dimension, information, and entropy, which all feature in the science of CA. And in Chapter 2, dynamical systems are mathematical by their very nature. However, in the remainder of the text, outbreaks of mathematics have been deliberately kept to a minimum. On the other hand, a more advanced and mathematical account of cellular automata can be found in the 2002 book by Ilachinski. An excellent pioneering work is, *Cellular Automata Machines,* by Tommaso Toffoli and Norman Margolus, both of whom have made important contributions to the field. Many of the topics mentioned herein, such as dynamical systems, chaos, artificial intelligence, and genetic algorithms, are only touched upon in the context of cellular automata, but many fine books are available on any of these. It is the author's intent that by being exposed to the tip-of-the-iceberg, the curious reader will be inspired to explore the vast wonderland below.

As much as possible, the author has tried to implement Marvin Minsky's dictum that "You don't understand anything until you learn it more than one way." Indeed, several important notions are presented from multiple points of view and throughout the text there are several recurrent themes, such as point attractors, periodic cycles, and chaos, among others.

One *caveat* concerning the applications of cellular automata. We aretex not making any claims that CA models are necessarily superior to other kinds of models or that they are even justified in every case. We are merely presenting them as one way of looking at the world which in some instances can be beneficial to the understanding of natural phenomena. At the very least, I think you will find them interesting. For those who already have a passing knowledge of the subject, I think you will discover some new things you have not seen before. Even if the entire universe is not one monolithic cellular automaton, as at least one scientist believes, the journey to understanding that point of view is well worth the price of admission.

Finally, I wish to thank Auckland University students Michael Brough, Peter Lane, Malcolm Walsh, as well as Dylan Hirsch-Shell of UCLA, and Nick Dudley Ward of Pattle Delamore Partners Ltd, who produced many of the figures from the CA models given in the text, and Samuel Dillon who produced the Rule 30 data encryption figures. Their assistance has been invaluable as their programming skills far exceed my own. I also wish to thank my daughter-in-law Yuka Schiff for many of the fine graphics and my friend Michael Parish for introducing me to the fascinating world of bees and Maeterlinck's classic monograph. A large debt of gratitude is owed to those who made valuable contributions to the manuscript in one form or another: Eshel Ben-Jacob, Michael Brough, Carlos Gershenson, Mario Giacobini, Dylan Hirsch-Shell, Jacques Mazoyer, Marc Ratkovic, Aaron Schiff, Birgitt Schönfisch, Dror Speiser, Guillaume Theyssier, and Jean-Baptiste Yunès. A very special thanks to Amy Hendrickson of TEXnology, Inc. for doing such an excellent and painstaking job laying up the text and images.

Several of the CA images were produced with the specialized software of Stephen Wolfram's *New Kind of Science Explorer*, which can be purchased from the website `http://www.wolframscience.com/`, and Mirek Wojtowicz's *MCell* program, which can be downloaded at his website `http://www.mirekw.com/ca/`. The latter also permits the real-time viewing of the evolution of some of the CA discussed in the text. Another interesting CA simulator is hosted by Michael Creutz at `http://quark.phy.bnl.gov/www/xtoys/xtoys.html`. Definitely have a look at Andy Wuensche's stunning website, `http://www.ddlab.com/`, and David Griffeath's Primordial Soup Kitchen, `http://psoup.math.wisc.edu/welcome.html`. An excellent CA resource website now edited by Tim Tyer is at `http://cafaq.com/`.

A website specifically for this text has been set up by John Wiley & Sons at ftp://ftp.wiley.com/public/sci_tech_med/cellular_automata/. Here there are further examples that have not been included in the text, Java applets, CA computer code, as well as a venue for interested readers to send in their own experimental contributions to the subject.

JOEL L. SCHIFF
University of Auckland

July 2007

Permissions

Figure 6.10	Reprinted from *Physica A* **284**, B. Luque and R. Solé, Lyapunov exponents in random Boolean networks, 33–45, Copyright © 2000, with permission from Elsevier.
Figure 6.12	Resnick, Mitchel, *Turtles, Termites, and Traffic Jams: Explorations in Massively Parallel Microworlds*, Figure 3.3. Copyright © 1994 Massachusetts Institute of Technology, by permission of The MIT Press.
Figure 6.13	Image courtesy Eshel Ben-Jacob, University of Tel Aviv.
Figure 6.16	Image from the website of Anahi Gajardo, `http://www.ing-mat.udec.cl/~anahi/langton/general.html`.
Figure 6.27	Redrawn from *Biosystems* **43**, M. Dorigo and L. M. Gambardella, Ant colonies for the travelling salesman problem, 73–81, Copyright © 1997, with permission from Elsevier.
Figure C.4	Image courtesy of Jacques Mazoyer, ENS-Lyon.
Figure C.6	Image courtesy of Birgit Schönfisch, University of Tübingen.
Figure C.10 and Figure C.12	Reprinted with permission of Christoph Hauert, from the *VirtualLabs* website `http://www.univie.ac.at/virtuallabs/`.
Figure C.11	Reprinted with permission of Serge Helfrich, from the website `http://prisonersdilemma.groenefee.nl/`.
Figure C.13	The code is courtesy of G. W. Flake (*The Computational Beauty of Nature*).
Figure C.14	Image generated by the *Wa-Tor* predator-prey program of Kovach Computing Services, `http://www.kovcomp.co.uk/`.
Figure C.15 and Figure C.16	Reprinted from *Chaos, Solitons & Fractals* **23**, C. Reiter, A local cellular model for snow, 1111–1119, Copyright ©2004, with permission from Elsevier.
Figure C.19	Reprinted with permission of Andy Wuensche, from the website `http://www.ddlab.org/`.
Cover art	Courtesy of Michael Brough, University of Auckland.

CHAPTER 1

PRELIMINARIES

Nothing shocks me. I'm a scientist.
–Harrison Ford (as Indiana Jones*)*

1.1 SELF-REPLICATING MACHINES

The origins of cellular automata can be traced back to mathematician John von Neumann's attempt to create a self-replicating machine. In 1948, von Neumann read a paper at the Hixon Symposium in Pasadena, California ("The general and logical theory of automata"), in which he outlined, among other things, a plan for a self-replicating machine (von Neumann actually refers to such a machine as an *automaton*). The question von Neumann addresses is this: "Can one build an aggregate out of such elements in such a manner that if it is put into a reservoir, in which float all these elements, each of which will at the end turn out to be another automaton exactly like the original one?" He then proceeds to outline the following argument to show that this is entirely feasible in principle.

One starts with a machine (universal constructor) A that has the ability to construct *any* other machine once it is furnished with a set of instructions denoted by *I*. Machine A is envisaged to float in the reservoir of liquid with all the necessary component parts that it requires for any particular construction. We now attach to our machine A another component called B that can make a copy of any instruction that is supplied to it. A final component, labeled C, von Neumann called the "control mechanism," which has the functions of initiating A to construct a machine as described by the instructions I and then cause B to make a copy of the instructions I and supply the copy of the instructions to the machine newly formed by A. Then the entire apparatus can be denoted by $M = A + B + C$.

To get things rolling, we furnish a machine M with a set of instructions for constructing itself, I_M, and call the resulting system M'. It is this machine, M', that is capable of replicating itself. For, C initiates the construction of M by A, it then has B make a copy of the instructions I_M and these are furnished to M to form the system M' once again. And so on.

It is the multiple use of the instruction set I_M that is crucial here. First, the instructions must be followed by A, second, they must be copied by B, and last, the copy must be attached to the machine constructed by A.

Overall, the copying mechanism is similar to the replication of living cells whereby the DNA (instructions) are first copied by cells preceding cell division. Interestingly, Christopher Langton [1986] comments: "Since he [von Neumann] was able to demonstrate that such a machine can exist, it becomes plausible that many, perhaps all, of the processes upon which life is based are algorithmically describable and that, therefore, life itself is achievable by machines." The study and implementation of these processes have become the domain of the newly emerging subject of *artificial life* and we shall encounter many instances of it throughout this book.

Von Neumann had now shown that a self-replicating machine was established in principle, but at the time of his lecture, he did not suggest how one could be implemented. The technology of the day simply was not capable of such an implementation.

According to received wisdom, it was the Polish-American mathematician Stanislaw Ulam who suggested to von Neumann that he should try constructing his self-replicating automaton using the conceptual framework of what are now known as cellular automata. The resulting system outlined in the early 1950s and later completed after von Neumann's death by Arthur Burks (see von Neumann [1966]) was a universal Turing machine embedded in a two-dimensional cellular lattice that had 29 states for each cell and a 5-cell neighborhood (now known as a von Neumann neighborhood) that required $\sim$200,000 cells. However, it was never actually implemented.

Exactly what is meant by *dimension* and *states* will be dealt with in the sequel.

A simpler 8-state self-replicating cellular automaton was created by Codd [1968] with some computer assistance. Then in 1984, Christopher Langton demonstrated self-reproduction in an 86-cell looped pathway using 8 states with a 5-cell neighborhood, which did not exhibit the feature of universal construction as did the von Neumann and Codd machines, but simply reproduced itself. Langton's loop has a construction arm attached to it and consists of an outer sheath of cells that remain in a fixed state and an inner sequence of "DNA" cells in various states that circulate around the loop (Fig. C.1, see color section). At the junction of the loop and arm, the DNA cells are replicated: One copy goes back around the loop and the other copy travels down the construction arm where it is translated at the tip of the arm spawning new growth.

Once a side of the offspring loop is fully generated, the growth pattern makes a left turn and propagates another side and so on until the offspring loop is complete. Then the connection between parent and offspring is severed (cutting the umbilical cord so to speak) and both parent and offspring propagate separate construction arms to begin the process anew (Fig. C.2, see color section).

Construction continues in this fashion with each new loop generating at least one new offspring. When a loop tries to extend an arm into a region already occupied, it will retract the arm and the DNA of that loop is erased and the loop becomes inert. The Langton loops will continue this replication process indefinitely expanding outward with time and filling the plane (Fig. C.3, see color section).

Although each loop contains the same DNA sequence, the number of times it can replicate itself will depend on the space available in its immediate environment.

A somewhat simpler self-replicating loop that dispenses with the outer sheath but also having eight states was constructed by Reggia *et al.* [1993]. A simple and brief proof of the existence of a self-replicating CA machine capable of universal computation was given by A. R. Smith [1991]. Another approach was taken by Morita and Imai [1997], who devised cellular configurations that were able to reproduce by self-inspection rather than from any stored self-description.

In the realm of actual self-reproducing machines, a primitive form was demonstrated by Roger Penrose (the well-known physicist) and his father Lionel back in 1957 using a set of flat shaped wooden blocks that produced copies of a particular coupling when the blocks were shaken in a box enclosure. In 2001, Greg Chirikjian of Johns Hopkins University developed a LEGO robot that drove around a track and assembled modules to make a copy of itself. Recently, Hod Lipson and colleagues (Zykov *et al.* [2005]) at Cornell University have created a self-replicating robot consisting of a tower of cubes that can swivel around and pick up other cubes and stack them to create another tower identical to itself. According to Lipson, this opens up the possibility of using robotic systems in future space travel that can repair themselves.

An overview of 50 years of research on self-replication can be found in the article by M. Sipper [1998], who also created an interactive self-replicator (Stauffer and Sipper [2002]). Certainly the notion of self-replication has proved enormously popular with the creators of computer viruses.

1.2 GRAND TURING MACHINES

In his seminal 1936 paper on computable numbers ("On computable numbers, with an application to the Entscheidungsproblem"), English genius Alan Turing discussed a very general type of computer that has become known as a Turing machine. This machine is theoretical in nature and today still finds applications in computer science. In the words of computer scientist Edward Fredkin, "It was a way to formalize all the things that a mathematician could do with a pencil and paper."

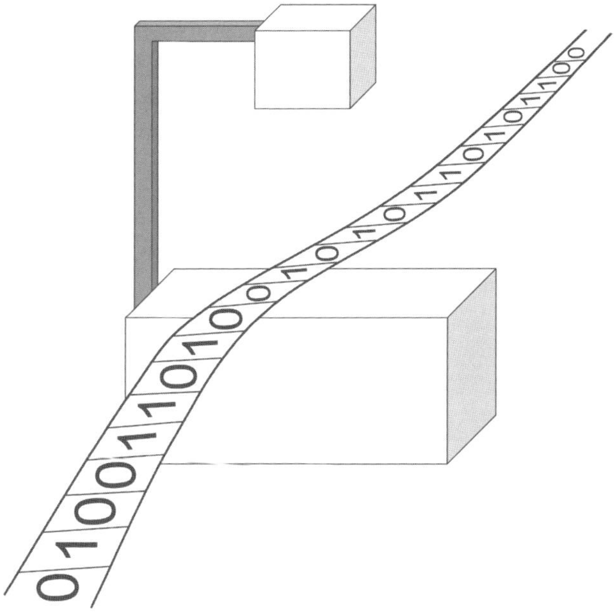

Figure 1.1 An idealized version of a Turing machine with the head reading the input of the tape below. The entire mechanism can move one square to either the right or left.

It can be thought of as a mechanical "head" that has an infinite strip of tape in both directions that lies beneath it. The head can both read and write onto the tape. The head is allowed to exist in a finite number of internal states, say k. The state of the head changes by interacting with the input from the tape beneath it. The tape is divided into an endless succession of square cells in which either a number 1 is written or the cell is blank, which we interpret as being the number 0. While the tape is infinite in length, there can only be a finite number of cells containing the number 1. The remaining cells must all be blank. The head reads just one such cell at a time — the one directly beneath it. Upon the head reading the value 0 or 1 of this cell, it replaces this value either with a 1 or 0 or with the same symbol. The head then moves one square either to the right or to the left (or not at all) and goes into one of its other allowable states. It then repeats the above sequence for each cycle of the machine. The new state of the head, the value the head gives to a particular cell, and the movement of the head left or right

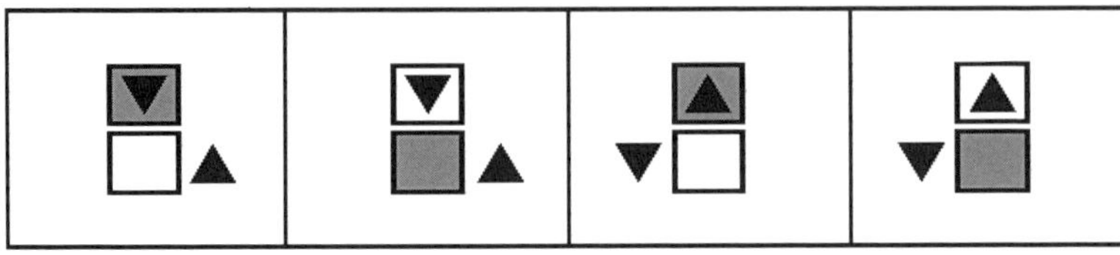

Figure 1.2 Graphical form of the Turing machine state transition function given in the text, denoting Head State 1 = ▼ and Head State 2 = ▲.

are all governed by some underlying set of instructions — the *state transition function*. There are also special starting and halting states. The output is written on a portion of the tape that can be read by an observer after a halt state has been reached (Fig. 1.1).

As an example, let us assume that the head has just two allowable states 1 and 2, with a state transition function given by the format

(**H**ead **S**tate, **C**ell **S**tate) $\longrightarrow$ (**H**ead **S**tate, **C**ell **S**tate, **M**ove)

specifically, say:

$$\begin{aligned}(1,1) &\longrightarrow (2,0,1);\ (1,0) \longrightarrow (2,1,1);\\ (2,1) &\longrightarrow (1,0,-1);\ (2,0) \longrightarrow (1,1,-1).\end{aligned}$$

These four rules could also be depicted in the graphical form of Fig. 1.2, where Head State 1 = ▼ and Head State 2 = ▲.

So, the first cell of Fig. 1.2 illustrates the first rule: $(1,1) \longrightarrow (2,0,1)$; meaning, if HS = 1 (i.e., ▼) and the head is on a cell with CS = 1 (gray), then HS becomes 2 (i.e., ▲), the CS is changed to 0 (white) and the head moves 1 cell to the right along the tape, and so forth with each rule. We can represent the evolution of this Turing machine by letting each step be depicted in a vertical downward direction with moves of $+1$ being to the right and -1 being to the left. Thus the above Turing machine would evolve as in Fig. 1.3 from an initial condition HS = 1, CS = 0.

Via the preceding rules, $(1,0) \longrightarrow (2,1,1)$, so that the HS is changed to 2, the initial cell state is changed to CS = 1, and the head moves one cell to the right. This state of affairs is denoted on the second line down, where we have the initial cell in state 1 (gray), and the head (in HS = 2) has moved over one cell to the right, over a cell in state 0 (white). Next

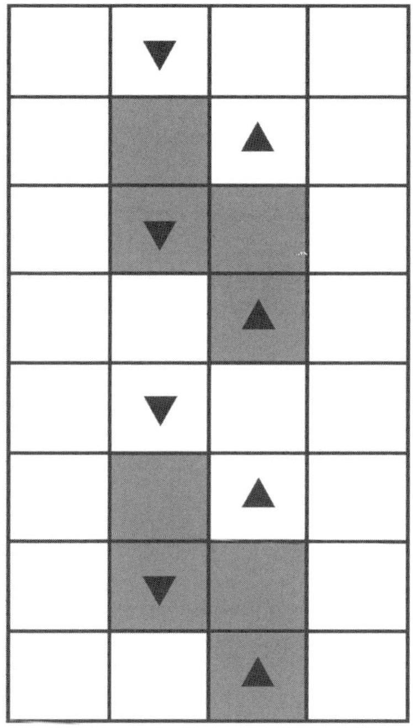

Figure 1.3 The time evolution (top to bottom) of the Turing machine example in the text starting with an initial condition HS = 1, CS = 0.

we find that $(2, 0) \longrightarrow (1, 1, -1)$, meaning that the HS now becomes 1, the CS becomes 1, and the head moves one cell to the left, directly over the initial cell (which was gray from before). Now HS = 1 and CS = 1, so we use $(1, 1) \longrightarrow (2, 0, 1)$, which alters the head to HS = 2, the cell state changes to CS = 0, and the head moves one cell to the right again (which also was gray from before). And merrily on the head goes, in this simple case never moving beyond the first two cells, and continually repeating its actions in a four-step cycle.

One question that one may ask about any particular Turing machine is, given a particular input, will the machine stop? This general question is known as the "halting problem" and it was shown by Turing himself that there is no way in principle to decide whether any particular Turing machine will stop or not. It may be noted that this result has ramifications to the whole field of mathematics itself.

Many mathematical questions can be posed in the language of whether or not a specific Turing machine will halt or not. For example, Opperman's conjecture states that for any integer $n > 1$, between the numbers

n^2 and $(n + 1)^2$ one can always find a prime number. For example, between $3^2 = 9$ and $4^2 = 16$ lies a prime, in fact two in this case, 11 and 13. As the integers n start to get larger and larger, the number of primes starts to thin out so there is always a possibility that there will not be sufficient numbers of primes to fit between every n^2 and $(n+1)^2$. Fortunately, n^2 and $(n + 1)^2$ spread out too. In spite of the elementary nature of this conjecture, no one has been able to prove or disprove it.

We could set up a Turing machine that would check to see if there was indeed a prime between every pair of numbers n^2 and $(n + 1)^2$, and we could instruct the machine to stop if for some pair a prime was not found. If the Turing machine does stop, then we have produced a counterexample to the Opperman conjecture and it is false. If somehow we knew that this Turing machine never stopped, then we would know that the Opperman conjecture was indeed true. But being able to solve the halting problem for this particular Turing machine is equivalent to determining the truth or falsity of the Opperman conjecture.

This means that there can be no general algorithm for deciding the truth or falsity of mathematical problems, which was the Entscheidungsproblem in the title of Turing's paper. This problem was first enunciated by the famous German mathematician, David Hilbert, at the 1900 International Congress of Mathematicians and was included in a list of 23 mathematical problems to be considered over the ensuing new century. This problem asked whether or not there was some general mechanical procedure that would be able to determine the truth or falsity of a large body of well-defined mathematical problems such as the Opperman conjecture. And as we have just seen, Turing showed this was not possible.

Turing machines with different sets of instructions are capable of doing different tasks, while there are some that are capable of emulating the performance of any other Turing machine. These are *universal Turing machines* and are said to be capable of performing *universal computation*. They can in fact do any calculation that is computable. Actually, the ubiquitous personal computer is effectively a universal Turing machine. Although a personal computer does not have an infinite storage capacity, it is essentially large enough to be considered so. Until recently, the simplest universal Turing machine was due to Marvin Minsky, who in 1962 produced a universal Turing machine with seven head states and four cell states. Based on the universality of Rule 110 (Section 3.5), Wolfram [2002] reduced this to just two head states and five cell states. In May 2007, Wolfram offered a $25,000 prize to anyone demonstrating that a particular two head state/three cell state Turing machine is universal (see `http://www.wolframscience.com/prizes/tm23/` for details).

Other work of Alan Turing's is encountered in Chapter 5 regarding reaction-diffusion equations, but we mention here just one further product of Turing's fertile mind — the *Turing test* — first proposed in 1950, although originally referred to by Turing as the *imitation game*. In this test, there is a human being in one room and a computer in another. Both communicate with those outside the room via some electronic device such as a teletype. An interrogator asks questions of the inhabitants of both rooms and tries to decide which is the human and which is the computer. If the interrogator, after a given period, is unable to determine which is which, the computer is said to have passed the Turing test. This is a very straightforward test of "machine intelligence" and in spite of many criticisms and objections over the years, it has stood the test of time very well. What it is not is a test of "machine consciousness" and this complex issue will be discussed further in Chapter 6.

Turing himself believed that in 50 years time an average interrogator after 5 minutes of questioning would have no more than a 70% chance of deciding which was the computer and which was the person. Nevertheless, as of the present date, no computer has even come close to passing the Turing test. In 1966, Joseph Weizenbaum of MIT created a computer program called ELIZA (after G. B. Shaw's Eliza Doolittle character of *Pygmalion* fame) that simulated the responses of a Rogerian psychotherapist. Some administrative staff members actually consulted ELIZA concerning their everyday problems and even the simple-minded Apple II version had many of this author's friends conversing intelligently with ELIZA for a few minutes. Readers are welcome to discuss their psychological problems with ELIZA at `http://www.manifestation.com/neurotoys/eliza.php3`.

1.3 REGISTER MACHINES

Computers basically perform operations on numbers that are stored in what are called *registers*. A (Minsky) register machine is just an idealization of the basic manipulations that are performed on the num-

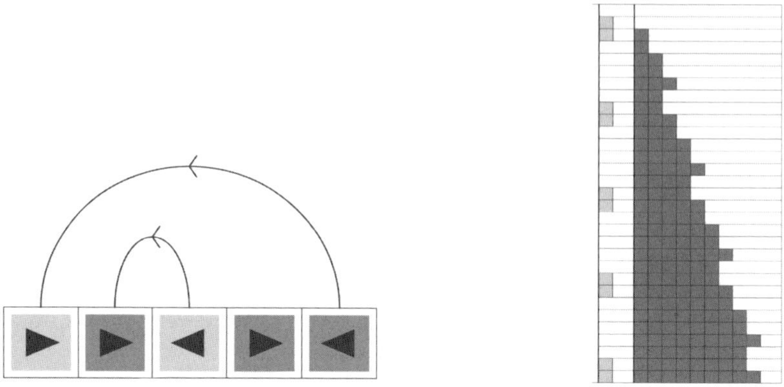

Figure 1.4 An example of a simple register machine with two registers and five instructions. Light gray indicates register 1 and dark gray register 2. An increment instruction is denoted by ▶ and a decrement by ◀ together with the path of the jump (arrowed). The output after each step is at the right with the first column displaying the contents of register 1 and the second column the contents of register 2. Both registers are initially empty (top line right) and the instructions begin with the incrementing of register 1 (first box left). Note the cyclic nature of the contents of register 1.

bers contained within the registers of a real computer. In order for the computer to perform a simple calculation such as the addition of two numbers, it must take the numbers stored in one of the registers and combine it with the number stored in another register. A register machine has three types of instructions: INC(rement), DEC(rement) and HALT. The INC instruction increments by 1 the number stored in a particular register and then the machine proceeds to process the next instruction. The DEC instruction has two components: it decrements by 1 the number stored in a particular register and then it will "jump" to another specifically designated instruction. But there is one *caveat* here. The number in a register cannot be less than zero, so if the value of zero is stored in a register and that register is to be decremented, then the instruction is ignored and the machine proceeds to the next instruction. Finally, the HALT instruction does simply that — it halts the operation of the register machine (Fig. 1.4).

Computer scientist Marvin Minsky [1967] showed that all one needs is just two registers to emulate a universal Turing machine.

The reader can create their own Turing machines and register machines and investigate their behavior using Wolfram's *NKS Explorer*.

1.4 LOGIC GATES

Logical statements are of the form: "April comes before May," "3 is less than 2," "All men are mortal" and are always considered to be either *true* or *false* and, as such, are given a truth value T (true) or F (false). This is a somewhat different philosophy from that normally assumed for statements in everyday conversation, but it is nevertheless the one adopted when discussing statements in a formal logical framework. We will denote declarative statements by capital letters such as:

P: All men are mortal.

Q: Socrates was a man.

Logical statements can be combined using connectives to form compound statements. The *conjunction* of statements P and Q is denoted by $P \wedge Q$. Consequently, if P and Q are as above, then $P \wedge Q$ is the statement "All men are mortal and Socrates was a man." The *disjunction* of P and Q is denoted by $P \vee Q$, and for the above P and Q, then $P \vee Q$ represents "All men are mortal or Socrates was a man," allowing for the possibility of both statements being true. The *negation* of a given statement P is denoted by $\neg P$. Thus if P represents the statement "All men are mortal," then $\neg P$ represents the statement "It is not the case that all men are mortal."

The truth value of the *Boolean relations* $\neg P$, $P \wedge Q$, and $P \vee Q$ is determined by the following principles and depends on the truth value of the statements P, Q according to the following truth table:

P	Q	$\neg P$	$P \wedge Q$	$P \vee Q$
T	T	F	T	T
T	F	F	F	T
F	T	T	F	T
F	F	T	F	F

Thus if the statement P: "All men are mortal" is true, then the statement $\neg P$: "It is not the case that all men are mortal" is false. Note that the conjunction $P \wedge Q$ is true only when both P and Q are true, whereas the disjunction $P \vee Q$ is true when either P or Q is true (or both are true).

At the heart of any computer system is the construction of the logic gates NOT, AND, OR that are simply implementations of the input and output of the preceding truth table. These gates regulate the flow of a current through the system where T symbolizes the state 1, or current

Figure 1.5 Line segment with two self-similar smaller segments and scale $r = \frac{1}{2}$.

on, and F symbolizes 0, or current off, as below (actually, high voltage and low voltage, but that is of no consequence).

P	Q	$\neg P$	$P \wedge Q$	$P \vee Q$
1	1	0	1	1
1	0	0	0	1
0	1	1	0	1
0	0	1	0	0

Thus, if we can somehow create the equivalent of the above logic gates in a particular environment, along with a mode of data storage, voilà, we have potentially all the power of a desktop computer at our disposal. In fact, we will do just this, more than once.

1.5 DIMENSION

Most people probably realize that we live in a three-dimensional space (well, actually four-dimensional, since we should consider *time* as another dimension) and that a plane is two-dimensional, a line one-dimensional, and a point zero-dimensional. These are the *Euclidean dimensions* of our everyday world. Various other dimensions have arisen in mathematics and in the sequel we will discuss several of them. Of special relevance are the types of dimension that have noninteger values used to study fractals and chaotic attractors.

Let us consider a line segment of fixed length and divide it in half. Then each half is a replica of the original except they have been scaled down by a factor of $r = \frac{1}{2}$. We call r the *scale* and observe that the number of *self-similar segments* N, which is two in this case, satisfies the relation $Nr^D = 1$, where $D = 1$ is the dimension of the line (Fig. 1.5).

In dimension 2, if we take a square and divide it up into $N = 4$ self-similar smaller sized squares (Fig. 1.6), the scale r of a side of a

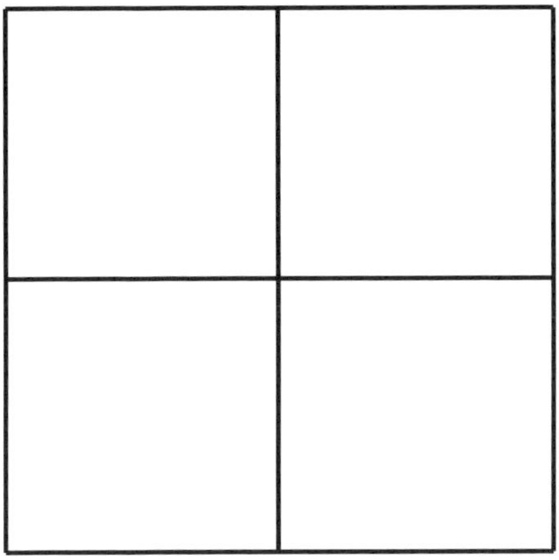

Figure 1.6 Square with four self-similar smaller sized squares and scale $r = \frac{1}{2}$.

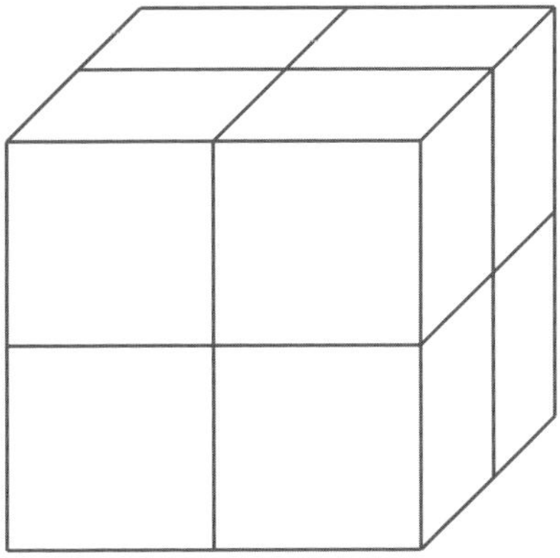

Figure 1.7 Cube with eight self-similar smaller sized cubes with scale $r = \frac{1}{2}$.

square is again $\frac{1}{2}$ and we find again that $4(\frac{1}{2})^2 = 1$, that is, $Nr^D = 1$, where here the dimension D equals 2.

Likewise for $D = 3$, a cube can be divided up into $N = 8$ self-similar but smaller sized cubes (Fig. 1.7) with the scaling factor $r = \frac{1}{2}$ and again $8(\frac{1}{2})^3 = 1$, so that $Nr^D = 1$ holds in all three cases. Since $N = r^{-D}$, we say that N obeys a *power law*.

Using the relation $N = r^{-D}$ and solving for D by taking the logarithm of both sides, we find that

$$D = \frac{\log N}{\log(1/r)}. \tag{1.1}$$

This is taken to be the definition of the *similarity dimension* and it agrees with the Euclidean dimension of a line, square, and cube.

Power laws appear in many areas of science and are important because they indicate relationships that are *scale invariant*, that is to say $y \propto x^d$ over a range of orders of magnitude of x-values. Thus we glean a little universal truth within the confines of a little universe. For example, metabolic processes, in mice, rabbits, dogs, goats, men, women, as well as elephants, are proportional to body weight to the power $\frac{3}{4}$ (Kleiber's law), whereas other biological functions for mammals and birds scale very closely to the body weight to the $\frac{1}{4}$ power (see for example the survey paper of J. H. Brown *et al.* [2002]). The power d may also be negative, such as in Zipf's law, which states that the frequency of a word's usage in English text is inversely proportional to the ranking of the word's frequency, that is to say, a word ranked nth place in the frequency table occurs $1/n$ times as often as the first ranked word. Let us not forget Newton's universal law of gravitation which states that the gravitational attraction between two bodies is inversely proportional to the square of the distance between them.

The World Wide Web is certainly not immune to power law mechanisms, which feature in the number of links, pages, and usage (see `http://www.hpl.hp.com/research/idl/papers/ranking/adamicglottometrics.pdf`). This has led to the general notion of *scale-free networks* (such as the Web), in which a few network nodes have a high degree of connecting links, thus referred to as hubs, with connectivity decreasing as the number of nodes increases, following some power law distribution, with $d < 0$. Such scale-free networks have very important properties such as tolerance to a random failure, since any random failure will most probably take place at a node of low connectivity.

The interesting thing about powers laws such as $y = ax^d$ (a constant) is that when you take the logarithm of both sides as we did in Equation 1.1, this results in

$$\log y = d \log x + \log a,$$

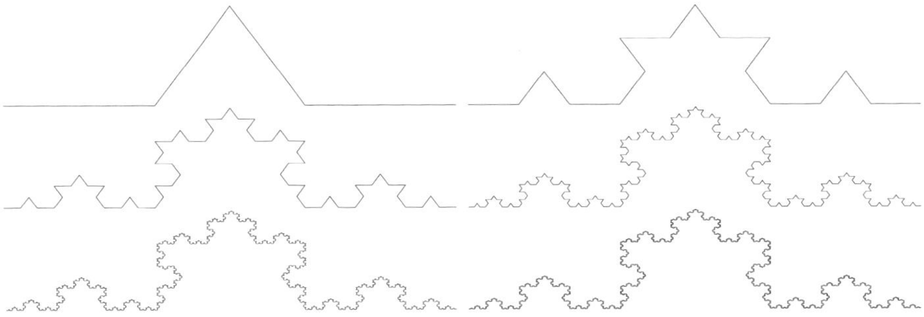

Figure 1.8 Starting from a straight-line segment, here are the first six steps leading to the production of the Koch curve.

which is nothing more than the equation of a straight line of the form $Y = mX + b$, whose slope has the value $m = d$. Conversely, suppose that we have data values y depending on values x and we plot a graph of $Y = \log y$ against $X = \log x$, known as a *log-log plot*. If the data so plotted "lies" on a straight line given by, $Y = mX + b$, then the slope m of that line is the power d, with x and y satisfying the power law relation $y = ax^d$. Of course d can have positive, negative, and noninteger values, as mentioned above.

Finding power laws in the natural world is one thing, but finding the mechanism giving rise to the power law is quite another (see Section 5.5.3 on the notion of self-organized criticality). Certain power laws connected with phase transitions have been shown to arise from fractal structures, in work that won Kenneth G. Wilson the 1982 Nobel Prize in Physics.

We will encounter the notion of a *phase transition* in various contexts throughout the sequel, and so let us simply define it as an abrupt change of a system from one state to another, brought about by varying a particular parameter of the system, as occurs for example, during freezing, melting, evaporation, condensation, sublimation, or superconductivity.

An example of a mathematical curve that does not have a proper integer dimension is the so-called *Koch curve*. It was given by Helge von Koch in 1904 as an example of a continuous curve that had no tangent at any point. It is constructed as follows. Starting with a line segment say of some unit length, replace the middle third of the line segment with a two-sided triangle having sides of length $\frac{1}{3}$ as indicated in Fig. 1.8.

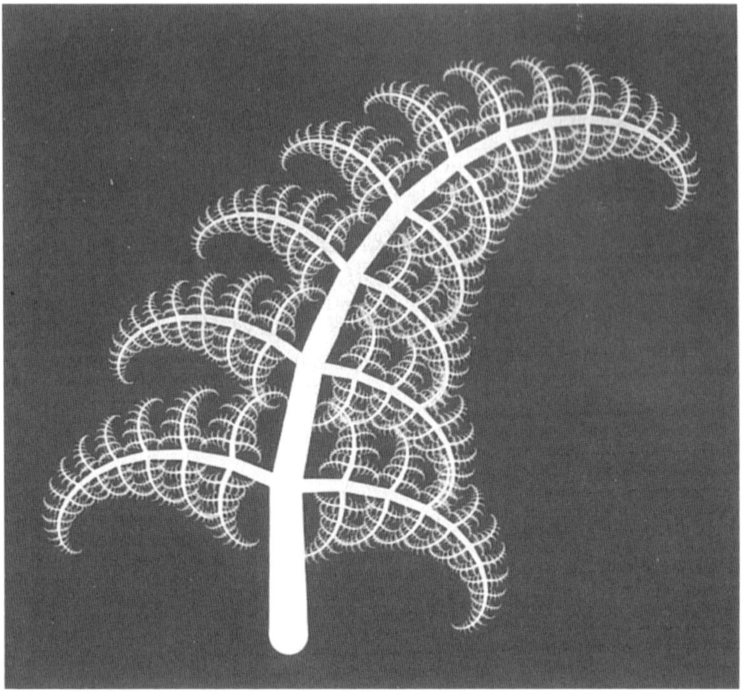

Figure 1.9 A fractal leaf form created using self-similarity.

We treat each resulting straight-line segment as above, replacing it with a two-sided (equilateral) triangle and continuing with this process *ad infinitum.* The limit of this process is the Koch curve that has no straight-line segments at all and no tangent at any point. Since the length of the curve grows by a factor of $\frac{4}{3}$ at each step, the limiting length is infinite. Moreover, since each line segment is replaced by $N = 4$ self-similar segments and the scale for these segments is $\frac{1}{3}$, the similarity dimension is $D = \log 4/\log 3 \simeq 1.262$. This dimension being greater than 1 gives some indication that the curve in some sense takes up more space than an ordinary line. Any figure that has a noninteger dimension is called a *fractal*, and the similarity dimension is also referred to as the *fractal dimension.*

Fractals have details at arbitrarily small scales, and when that detail is in the repeated form of the whole, they are *self-similar*. This notion has been very useful in the computer generation of many artificial life forms (cf., e.g., Mandelbrot [1982], Oppenheimer [1989]), as in the fern leaf (Fig. 1.9).

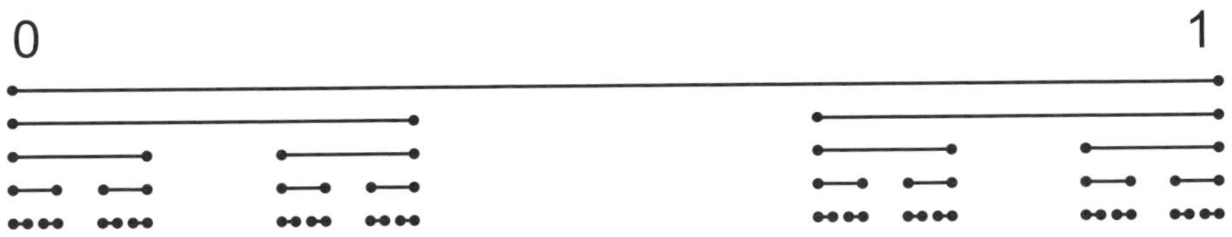

Figure 1.10 The construction of the Cantor set whereby the middle third of each remaining line segment is removed.

Another mathematical construction that is rather fascinating and also has a noninteger similarity dimension is the famous *Cantor set*, due to the German mathematician Georg Cantor. We start with a line segment of length 1 and remove the middle third of it, $(\frac{1}{3}, \frac{2}{3})$, leaving the two closed intervals, $[0, \frac{1}{3}]$ and $[\frac{2}{3}, 1]$.

Next, we remove the middle third of these two intervals, that is, $(\frac{1}{9}, \frac{2}{9})$ and $(\frac{7}{9}, \frac{8}{9})$. This leaves four closed intervals similar to the original, and we again remove the middle third of each, and continue on indefinitely as indicated in Fig. 1.10. The limiting set, which consists of nothing but the endpoints of intervals, but no interval itself, is the Cantor set. Interestingly, if we total up the lengths of the line segments that were removed, we obtain

$$\begin{aligned} \frac{1}{3} + \frac{2}{9} + \frac{4}{27} + \cdots &= \left(\frac{1}{3}\right)\left(1 + \frac{2}{3} + \left(\frac{2}{3}\right)^2 + \cdots\right) \\ &= \left(\frac{1}{3}\right)\left(\frac{1}{1 - \frac{2}{3}}\right) = 1, \end{aligned}$$

since the infinite sum is a geometric series. Thus we have removed from the unit interval an amount totaling 1 in length but have in fact left behind an uncountable number of points in the process! As for the dimension of Cantor's barely noticeable set of points, note that $r = \frac{1}{3}$ and $N = 2$, so that the similarity dimension is $D = \log 2/\log 3 \simeq 0.631$. Even though the Cantor set has the same number of points as a solid line, its dimension is less than 1 because in some sense we have removed so much. On the other hand, a finite set of points, or even a countable set of points like the integers, has similarity dimension zero. You can see why mathematicians just love the Cantor set.

Our final example will be encountered again when we talk about one-dimensional cellular automata and is the so-called Sierpiński triangle.

Figure 1.11 The Sierpiński triangle is the limiting case of the above figure.

We start with an equilateral triangle and remove an inverted version of a similar triangle having scale $r = \frac{1}{2}$ as in Fig. 1.11. This results in $N = 3$ similar triangles from which we again remove an inverted version of a similar triangle having scale $r = \frac{1}{2}$. Continuing in this way results in the Sierpiński triangle having similarity dimension $D = \log 3 / \log 2 \simeq 1.585$.

1.5.1 Kolmogorov Dimension

This is also known as the *box-counting dimension* and consists of a grid of boxes that covers the figure to be measured. The *log* of the number of boxes, $N(b)$, required to cover the figure, is plotted against the *log* of the length of the side of a box, b, for various box sizes. Again, a power law relation would arise from a straight line representing the log-log data points and the line's slope would give the more sophisticated sounding *Kolmogorov dimension,* D_K, of the figure (see Fig. 1.12). Without going into the mathematical details, an equivalent way to get our hands on the slope is from the formula

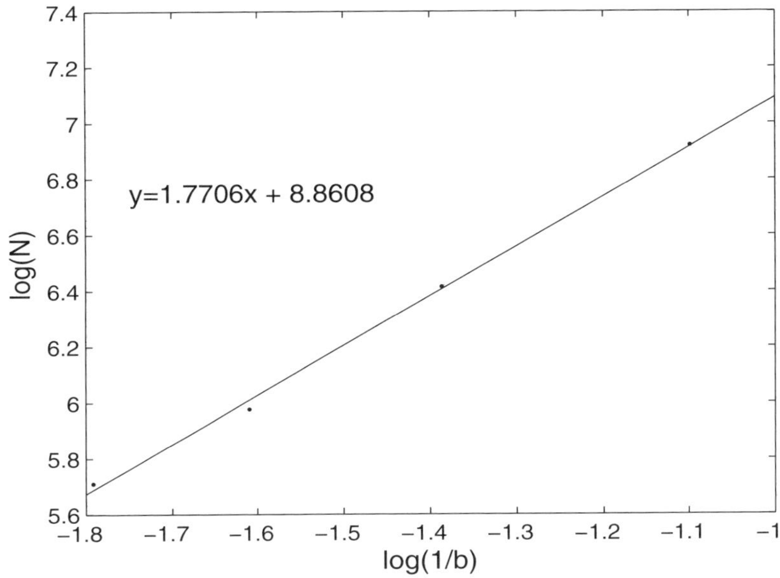

Figure 1.12 This is a plot of the log of the number of boxes needed to cover the Sierpiński triangle *vs.* the log of the box size in mm. The slope of the straight line gives the Kolmogorov dimension of 1.7706, in reasonably close agreement with the similarity dimension of 1.585 determined previously.

$$D_K = \lim_{b \to 0} \frac{\log N(b)}{\log(1/b)}.$$

If we consider the Cantor set and cover it with three bins, $[0, \frac{1}{3}]$, $[\frac{1}{3}, \frac{2}{3}]$, and $[\frac{2}{3}, 1]$, then we see that only two of these contain points of the Cantor set since the middle third is void by definition and we can count the endpoints $\frac{1}{3}$ and $\frac{2}{3}$ as belonging to the first and third bins with the other points of the set. As $b = \frac{1}{3}$, we see that $\log N(b)/\log(1/b) = \log 2/\log 3$, and in general, taking the number of bins to be 3^n, we find that exactly 2^n boxes contain points of the Cantor set, so that

$$\frac{\log N(b)}{\log(1/b)} = \frac{\log 2^n}{\log 3^n} = \frac{\log 2}{\log 3} \simeq 0.631,$$

which is just the similarity dimension that was determined previously.

1.6 INFORMATION AND ENTROPY

There is some kind of mystique about entropy.

— *Jean Bricmont*

The mathematical treatment of information began with the work of Claude Shannon [1948] and is based on certain probabilities. If an event is unlikely to occur (i.e., has low probability), then the amount of information obtained about its occurrence is relatively high, whereas if the event has a high probability of occurring, then the amount of information regarding its occurrence is relatively low. If the probability of the event in question is denoted by P, then the information it conveyed by its occurrence is defined to be

$$I = \log_2 (1/P)\,.$$

Although we have taken base 2 here, it can vary depending on the context, but for our purposes we will use base 2. For example, information in a computer is stored as sequences of 0's and 1's, offering two possibilities for each digit. Each digit, which has a 1 in 2 chance of appearing thus holds $\log_2(2) = 1$ *bit* (*binary digit*) of information. Likewise, if you toss a coin and a heads turns up, then 1 bit of information is conveyed. Or, if you roll a six-sided die with each side having a probability of $r = \frac{1}{6}$ of turning up and you roll a 5, then the information conveyed equals $\log_2(6) = 2.59$ bits of information. In fact, when any face turns up upon rolling the die, 2.59 bits of information are conveyed.

When we are dealing with events that have unequal probabilities, then the total amount of information is a weighted sum with the weight given to each contribution determined by the probability of its occurrence. For instance, if P_1, P_2,..., P_N are the probabilities of N individual outcomes subject to $\sum_{i=1}^{n} P_i = 1$, then the *information* I of the entire system is given by

$$I = \sum_{i=1}^{N} P_i \log_2 (1/P_i)\,. \tag{1.2}$$

I is in general positive but can be zero if all but one of the probabilities are zero and the remaining one equals 1. I becomes a maximum when all the probabilities P_i are equal to $1/N$ and the maximum value of I is then the sum of N quantities of $\frac{1}{N}\log N$, giving $I_{\max} = \log N$.

This expression for information is formally related to the concept of entropy that originated from the work of Rudolf Clausius in 1865 as a measure of the amount of irreversible heat loss of a system and who formulated the Second Law of Thermodynamics as the notion that *the thermodynamic entropy of a closed system increases to a maximum.* This has led to the belief in the eventual so-called heat-death of the universe, whereby the universe itself will reach a state of maximum entropy where all matter is cold and disordered. The Second Law was formalized by Boltzmann in 1872, who described entropy (H) by the mathematical formulation

$$H = -K \sum_{i=1}^{N} P_i \log_e P_i.$$

Boltzmann was interested in describing the dynamics of a gas and here the P_i represent the probabilities of gas molecules being in one of N particular states and K is the Boltzmann constant. With $K = 1$ we formally have the information I (except for a change of base) formulated by Shannon, and in fact, I is also referred to as *entropy* or *information entropy.* The concept of entropy has also been associated with disorder and randomness whereby high entropy is equivalent to a high degree of disorder (randomness) and low entropy is equivalent to a low level of disorder (non-randomness). In the sequel we will invoke the notion of entropy in this sense. It is sometimes claimed that biological systems violate the Second Law of Thermodynamics because they are ordered systems that arose from less ordered ones, but this is incorrect as biological systems are open to the environment and therefore cannot be considered closed. The Second Law describes *irreversible* processes (see Section 3.7) that increase in entropy along an arrow of time. Entropy is embraced by many other scientific disciplines, and has had a colorful and controversial history which continues to this day. According to Denbigh and Denbigh [1985], "A number of scientists and information theorists have maintained that entropy is a subjective concept and is a measure of human ignorance. Such a view, if it is valid, would create some profound philosophical problems and would tend to undermine the objectivity of the scientific enterprise as a whole." However, the authors conclude that "*thermodynamic* entropy is fully objective... and the same must apply to any other 'entropy' which is used as a surrogate."

There is also an *information dimension* associated with the information I. Let us consider the two-dimensional case in which the figure to be measured is overlayed with a grid of n same-sized cells or bins as in the Kolmogorov dimension. To compute the information I, each probability P_i represents the number of figure elements in the ith box, divided by the total number of figure elements, say N. This is just the *relative frequency* of figure elements occuring in the ith box. Then I is computed according to the equation

$$I = \sum_{i=1}^{N} P_i \log_2(1/P_i).$$

If a particular box has no figure elements in it, then corresponding probability P_i is zero and there is no contribution to the information I in this case. Evidence suggests that I increases linearly as a function of $\log(1/b)$, where b is the length of the side of each bin over a suitable range of values. Of course if the bin size becomes too large and the entire figure fits into one bin, then $P = N/N = 1$ and hence $I = P\log_2(1/P) = 0$. On the other hand, if the bin size becomes too small, then at most one figure element lies in any given box and each nonzero P_i equals $1/N$. If we just take our sum over the bins that are nonempty, we see that there are now N of these and I is equal to $N \times \frac{1}{N}\log_2 N$ giving $I_{\max} = \log_2 N$, the maximum value for I. Any further decrease in bin size gives no further increase in information but only the same value for I.

The information dimension is given mathematically by

$$D_I = \lim_{b \to 0} \frac{I}{\log_2(1/b)}. \tag{1.3}$$

To actually compute D_I, one can use the usual trick of graphing I against $\log_2(1/b)$ and the slope of the resulting straight line gives the required dimension (see Fig. 1.13).

Note that in the event when all the probabilities are equal, namely each $P_i = 1/N$, then we found that the information became a maximum and had the value $I = \log_2 N$. Substituting this into D_I above gives

$$D_I = \lim_{b \to 0} \frac{\log_2 N}{\log_2(1/b)},$$

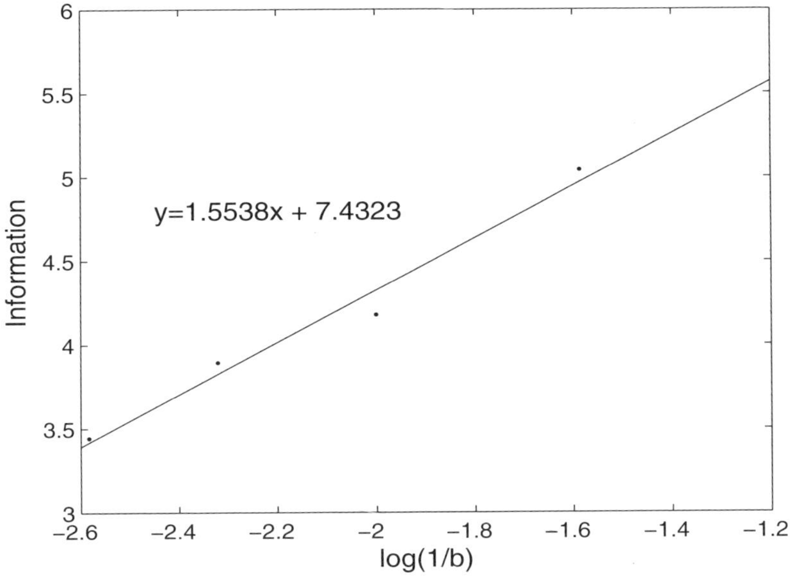

Figure 1.13 Here we have experimentally computed the information dimension for the Sierpiński triangle by the method explained in the text. The value of 1.5538 is in close agreement with the mathematically determined value of 1.585.

which bears a striking resemblance to the Kolmogorov dimension, D_K. *Plus ça change, plus c'est la même chose.* We will encounter several instances of this wonderful French expression.

1.7 RANDOMNESS

In a general sense, we say that something is random if it has no regular pattern or it is unpredictable. In a more technical sense, an infinitely long sequence is random if its Shannon information I is infinite. But for finite sequences which exist in the real world, many tests have been developed that verify whether or not the sequence in question shares certain statistical properties with a hypothetical random sequence.

Some randomness like that displayed in coin tossing or dice throwing is due to an exquisite sensitivity to the initial conditions that produced the outcome. Moreover, it has been known since the early twentieth century that physics at the atomic level is governed by a degree of randomness such as in the radioactive decay of the atoms of certain elements. Using the random nature of quantum physics, a commercial quantum random number generator has been recently developed (called *Quantis*) that fires photons (light particles) at a semitransparent mirror in which 50% are either reflected or transmitted with two detectors picking up the respective signals to generate a random sequence of 0's and 1's. To many scientists (although not every scientist is convinced about *Quantis*), the preceding examples are considered the only true form of randomness, because the outcome for any given event is not predictable or even repeatable. On the other hand, some deterministic randomness like the values of the digits of $\pi = 3.141592653...$ or $e = 2.718281828...$ or $\sqrt{2} = 1.414213562...$ and some of the one-dimensional cellular automata of Chapter 3 are computable from a simple algorithm. Or, as we note in the next chapter, the random behavior in the evolution of the logistic equation when $a = 3$ is predictable on short time scales but appears to be random in the long term. However, this sort of deterministic randomness is considered only *pseudorandom* by many because the output is always computable. Even the random numbers generated by computers are only pseudorandom since a computer is a deterministic system. According to my colleague, computer scientist Cristian Calude, given a sufficient amount of time and money, he would gladly compute the Nth digit of any of the above three numbers for any value of N. True random number generators like the Quantis are useful in cryptography and the scientific modelling of certain complex systems. However, for our purposes it will not be necessary to make any distinction between a truly random sequence and a pseudorandom one.

One topic that is intimately connected with cellular automata and shares many of its features is that of dynamical systems, which we take up in the next chapter.

CHAPTER 2

DYNAMICAL SYSTEMS

... it may happen that small differences in the initial conditions produce very great ones in the final phenomena.

— *Henri Poincaré – essay* Science and Method

One of the exciting new fields to arise out of twentieh century mathematics is that of dynamical systems. Topics like chaos and strange attractors have become nearly household words even if most people do not know their precise meaning. Dynamical systems arise in the study of fluid flow, population genetics, ecology, and many other diverse fields where one seeks to model the change in behavior of a system over time. Several of the global features of dynamical systems such as attractors and periodicity over discrete time intervals also occur in cellular automata, and thus it is worthwhile to have a working knowledge of the former. Indeed, cellular automata are dynamical systems in which space and time are discrete entities. We present here a brief summary of the salient features of dynamical systems and for the in-

Cellular Automata: A Discrete View of the World. By Joel L. Schiff terested reader there are many fine texts on the subject (e.g., Devaney [1989], Elaydi [2000], Sandefur [1990], Williams [2001]).

Certainly, one of the oldest and most notable examples of a dynamical system is the *Fibonacci sequence*,

$$1, 1, 2, 3, 5, 8, 13, ...,$$

where each sucessive Fibonacci number, let us call it, F_n, takes the value given by the sum of the two preceding numbers, F_{n-1} and F_{n-2}. So we have

$$F_n = F_{n-1} + F_{n-2}$$

for $n = 3, 4, 5, ...$ with $F_1 = F_2 = 1$. Leonardo Fibonacci of Pisa was interested in determining how many rabbit pairs one would have after n months, starting with just a single male/female pair. He made the assumption that newborn rabbits would begin breeding after two months had elapsed, producing another new male/female pair every month, and that no rabbits died. So, to calculate the rabbit pair population at the nth month, again denoted by F_n, we note that the preceding month's rabbit pair population was F_{n-1}, and of those, only F_{n-2} pairs are capable of reproducing (the same number of) new pairs. Thus, F_n is the sum of F_{n-1} and F_{n-2}, which is the Fibonacci *recurrence relation* above. Of course, Fibonacci's assumptions are not very biologically sound, but the sequence is fascinating and Fibonacci numbers turn up in many aspects of Nature, such as in the number of petals on various flowers.

Although Fibonacci's work dates back to 1202, the sequence itself was known to Indian scholars much earlier. The Fibonacci sequence has many interesting mathematical properties, one of them being that the quotient F_n/F_{n-1} tends to the value of the "golden ratio," $\varphi = (1 + \sqrt{5})/2$, as the value of n becomes infinite.

Let us now suppose that, instead of rabbits, we are interested in the size of some particular seasonally breeding insect population whose generations do not overlap. Denote the initial population size by x_0, after the first year by x_1, after the second year by x_2, and so on, with the population after the *n*th year denoted by x_n. Since the population size in any given year determines what it will be in the following year, we can write this deterministic feature of the population size in the form of a *dynamical system*:

$$x_{n+1} = f(x_n), \quad n = 0, 1, 2, 3, ...,$$

where f signifies some mathematical function. The set of all the iterations x_n, $n = 0, 1, 2, 3, ...$ is known as the *orbit* or *trajectory* of the dynamical system. So, for example, if the population size happens to triple from one generation to the next, we would have the relation

$$x_{n+1} = 3x_n, \quad n = 0, 1, 2, 3, ...,$$

to describe the changing population size from one generation to the next. More generally, a linear growth model would be

$$x_{n+1} = ax_n, \quad n = 0, 1, 2, 3, ...,$$

where if $a = 1$ the population remains static, and if $a > 1$ or $a < 1$ the population at each generation either grows or declines at a constant rate respectively.

More realistic equations taken from the biological literature are

$$x_{n+1} = (1 + a)x_n - bx_n^2, \quad n = 0, 1, 2, 3, ..., \tag{2.1}$$

and

$$x_{n+1} = x_n e^{[c(1-x_n)]},$$

where the parameters a, b, and c are constants which have values restricted to a certain range that allow one to "tune" the behavior of the dynamical system by varying these values appropriately. These parameters in some sense reflect the environmental conditions that the dynamical system is operating in. Equation 2.1 is referred to as the *logistic equation* and is used to model population growth, with the term b functioning as a "damping factor" because the term $-bx_n^2$ inhibits the growth of the population. These equations are *nonlinear* meaning that the output is not proportional to the input as in the preceding system. We will consider a special case of the logistic equation for which $b = a$, namely:

$$x_{n+1} = (1 + a)x_n - ax_n^2, \quad n = 0, 1, 2, 3, ...,$$

where a is a parameter in the range $0 < a < 3$. A surprising degree of complex dynamics is locked within this very simple equation and it will serve to model all the characteristic features that we wish to exhibit.

First however, I cannot resist presenting an example of this model that features in James T. Sandefur's book [1990]. Suppose that x_n

represents the fraction of those people on Earth who believe in alien beings at time period n, so that the quantity $1 - x_n$ is the fraction of nonbelievers. Furthermore, assume that in each time period the believers are able to win over a proportion of nonbelievers (or *vice versa*) and that the proportion depends on the interaction (i.e., product) of the two groups: $x_n(1 - x_n)$. This now can be represented as the dynamical system

$$x_{n+1} = x_n + ax_n(1 - x_n),$$

which is just the logistic equation above. The parameter a can be positive, indicating that the belief in aliens is increasing, whereas $a < 0$ indicates belief is ebbing.

In order to get this system underway, let us take the value $a = 1$ and some initial value, say $x_0 = 0.25$, and compute the subsequent iterations x_n. So, for example,

$$x_1 = 2x_0 - x_0^2 = 2(0.25) - (0.25)^2 = 0.4375,$$

and furthermore $x_2 = 0.6836, x_3 = 0.8999, x_4 = 0.9900, x_5 = 0.9999,$ $x_6 = 1.0000$, $x_7 = 1.0000$, $x_8 = 1.0000, ...,$ with all subsequent iterations having the value $x_n = 1.0000$. These values represent the orbit of the dynamical system and in this instance we refer to the value 1 as being an *attractor*. Taking some other initial value, say $x_0 = 0.5$, we find that the iterations $x_1 = 0.7500$, $x_2 = 0.9375$, $x_3 = 0.9961$, $x_4 = 1.0000$, $x_5 = 1.0000$, $x_6 = 1.0000...$ are once again attracted to the value 1. In fact, over the entire permitted range of initial values where our dynamical system operates, the iterations will be attracted to the value 1. What's more, by allowing the parameter a to vary over the range $0 < a < 2$, we find that the value $x = 1$ is still an attractor for any sequence of iterations (Fig. 2.1). The set of points that evolve to an attractor are said to lie in its *basin of attraction*.

To examine the nature of the attractor further, let us write our dynamical system as an ordinary function, namely

$$f(x) = (1 + a)x - ax^2,$$

so that

$$f(x_n) = (1 + a)x_n - ax_n^2 = x_{n+1}.$$

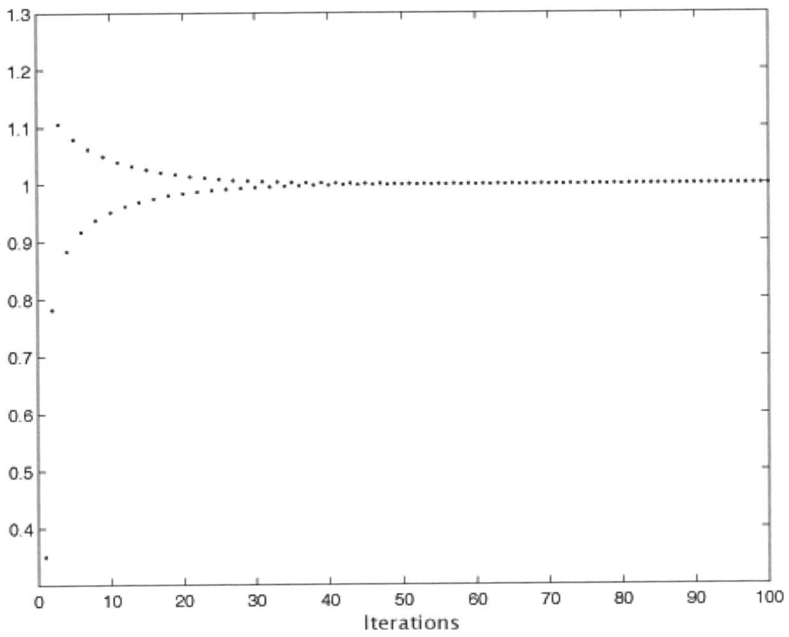

Figure 2.1 Here we have taken $a = 1.9$ and initial value $x_0 = 0.35$. The iterations are seen to converge to the value 1, the attractive fixed point.

With $a = 1$ the preceding attractor $x = 1$ has the property that $f(1) = 1$. Any value x that has the property $f(x) = x$ is called a *fixed point* of the function f. Are there any other fixed points of our function? Setting

$$f(x) = (1 + a)x - ax^2 = x$$

and solving for x, we find that $x = 0, 1$ are the only fixed points. But these two fixed points are qualitatively different. While $x = 1$ is a fixed-point attractor (also called a *stable* fixed point), the point $x = 0$ is a fixed-point *repeller* (or *unstable* fixed point). This means that the iterates will always be repulsed from the vicinity of $x = 0$, no matter how close to it they may start (Fig. 2.2).

There is a *derivative test* we can perform on the function f which can tell us *a priori* what type of fixed point we have. In particular, if $|f'(x)| < 1$ at a fixed point x, then it is an attractor, and if $|f'(x)| > 1$, the fixed point is a repeller. Note that we have $f'(x) = 1 + a - 2ax$,

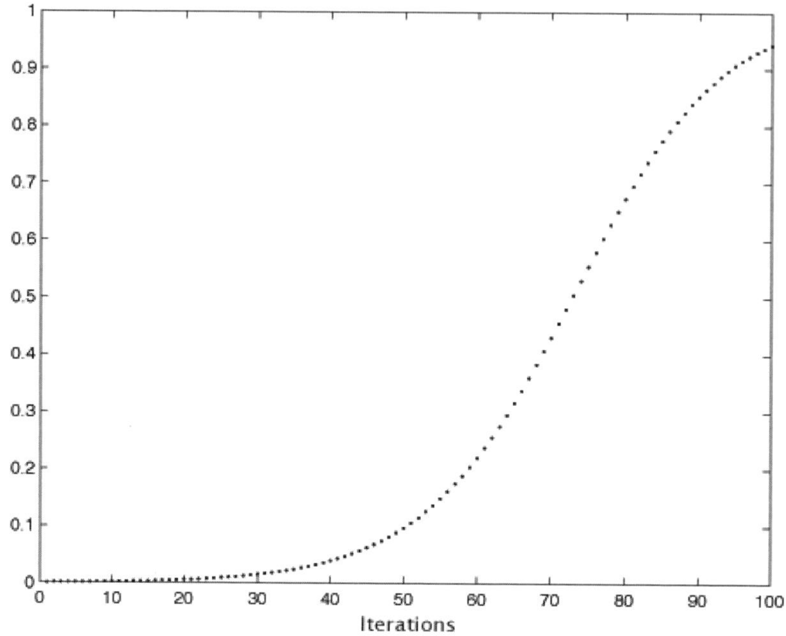

Figure 2.2 Here we have taken an initial value to be $x_0 = 0.001$ showing the repulsive nature of the value $x = 0$ and the attractive nature of $x = 1$ whenever $0 < a < 2$.

so that whenever $0 < a < 2$, $|f'(0)| = 1 + a > 1$, indicating a repeller at $x = 0$, and $|f'(1)| = |1 - a| < 1$, indicating an attractor at $x = 1$.

For the region $2 < a < 2.449$, we obtain an *attractor of period 2*, or *2-cycle*, with the dynamical system oscillating between two fixed values as in Fig. 2.3. Even iterates will be attracted to one of the values and odd iterates to the other. Periodic behavior is also found in certain cellular automata as will be pointed out in the sequel, although it is not such a significant feature as it is in dynamical systems.

Gradually increasing the parameter a, the 2-cycle is followed by a 4-cycle (Fig. 2.4), 8-cycle, and so on. Each of these cycles originates for a in some increasingly narrow band of values.

We can illustrate the behavior of our dynamical system by plotting the parameter value along one axis and plotting the values of the attractor along the other. At certain critical values of the parameter a, the dynamical system undergoes a bifurcation in the values generated and this produces a *bifurcation diagram* as depicted in Fig. 2.5. One such

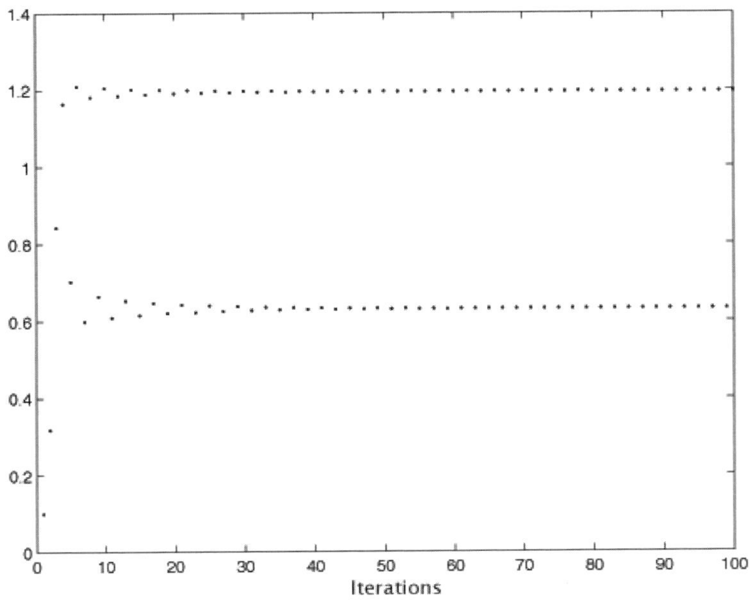

Figure 2.3 A 2-cycle for the parameter value $a = 2.42$ with the dynamical system oscillating between two fixed values, 0.6317... and 1.1947... .

value was $a = 2$, whereby the dynamical system went from having a stable attractor (at $x = 1$) when $0 < a < 2$ to a 2-cycle for values of a slightly larger than 2.

When $2.57 < a \leq 3$, the period doubling behavior of our dynamical system breaks down and we enter the *chaotic regime*. Here there are infinitely many initial values x_0 that yield *aperiodic orbits*, that is, the iterates x_n wander erratically over a fixed interval. Of course, these can be difficult to distinguish from cycles with very long periods. For example, there is an odd period 3-cycle at $a = 2.840$ (Fig. 2.6), which happens to exhibit its own period doubling (a 6-cycle at $a = 2.845$ and a 12-cycle at $a = 2.848$ were found). Even period cycles also turn up in this region, for example, at $a = 2.9605$ we have a return to a 4-cycle. All of these periodic cycles and infinitely more others exist within narrow windows of the parameter space.

At the end of our parameter interval when $a = 3$ (and at other values as well), our dynamical system effectively becomes a deterministic random number generator, that is, a pseudorandom number generator.

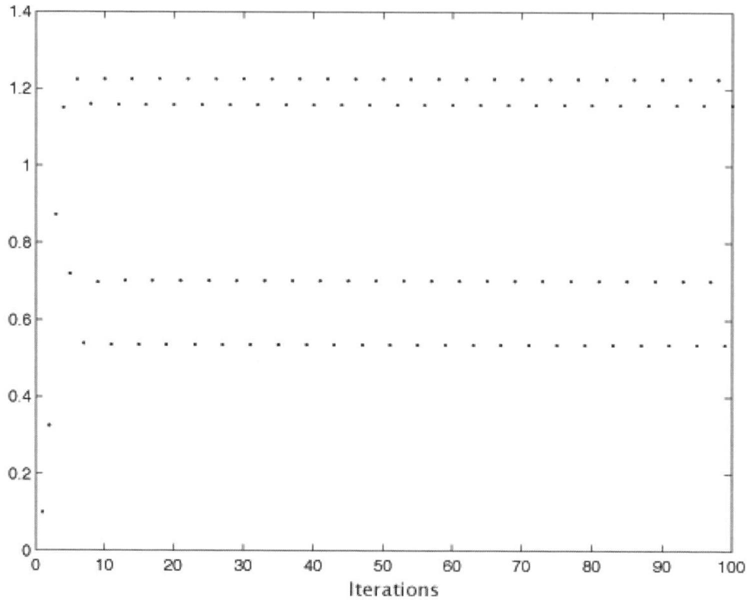

Figure 2.4 Here we have an attractor of period 4, or 4-cycle with $a = 2.5$ showing how the system oscillates between four distinct values.

At this point, we have what is commonly called *chaos*. One main feature (but not the only feature) of chaos is what is called *sensitivity to initial conditions*, that is, if we start off with two different initial values, say x_0 and y_0, no matter how close together, the resulting iterations x_n and y_n will eventually diverge away from each other. This feature makes long term predictions about the dynamical system impossible but does allow for short term predictions since initial values of the respective trajectories $x_1, x_2, \ldots$ and $y_1, y_2, \ldots$ will remain close. This sensitivity is nicely illustrated with the following experiment, where the initial values x_0 and y_0 only differ in the 12th decimal place by one digit (one part in a trillion)! Yet within 40 iterations, the respective iterations x_n and y_n are already widely divergent.

$$\begin{aligned} x_0 &= 0.123456789012 \\ x_1 &= 0.448102419788 \\ x_2 &= 1.190022343293 \\ &\vdots \\ x_{40} &\doteq 0.508890113704 \end{aligned} \qquad \begin{aligned} x_0 &= 0.123456789011 \\ x_1 &= 0.448102419785 \\ x_2 &= 1.190022343288 \\ &\vdots \\ x_{40} &\doteq 1.006900670500. \end{aligned}$$

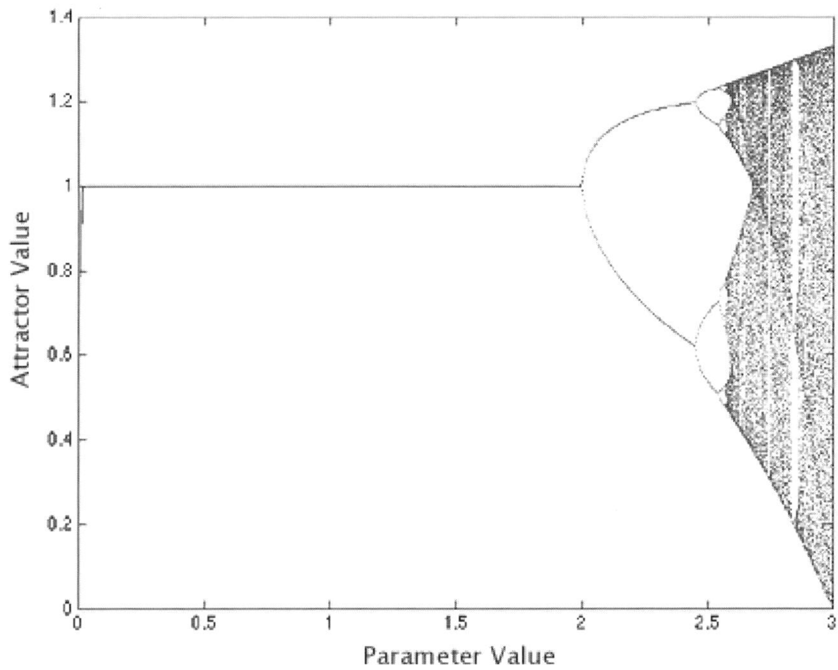

Figure 2.5 This diagram demonstrates how the values of the attractor of the dynamical system change as the parameter value a is varied and illustrates the common phenomenon of *bifurcation* and *period doubling* whereby a hierachy of 2^n-cycles are generated. The range of values of a over which the 2^n-cycle remains an attractor (i.e., stable) gradually diminishes.

The corresponding iterations remain close at first but gradually start to diverge, so that by the 40th iteration they are widely divergent and continue so thereafter. This extreme sensitivity to initial conditions is typical of a dynamical system in a chaotic state. An analogous situation arises in Class III cellular automata (see Section 3.8). It should be noted that when a dynamical system is not in a chaotic mode, say for a positive value of a that is less than 3 in the above system, then small differences in the input only result in small differences in the output.

A chaotic dynamical system has the feature that in practice long term predictions are impossible since one can never measure initial conditions with infinite precision. Therefore, as in the example above, the long term future of the iterations is in the hands of the gods. However,

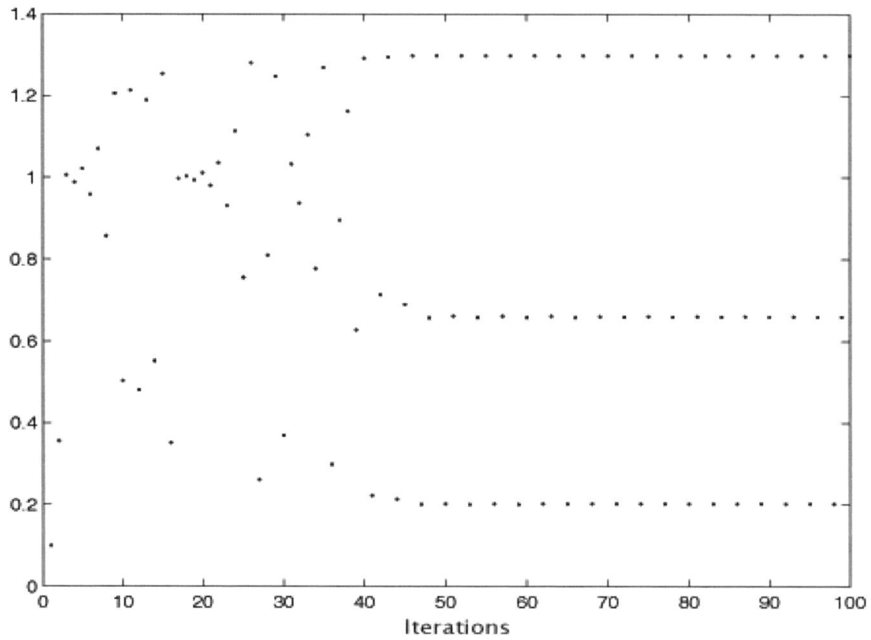

Figure 2.6 In the chaotic regime we can find even- and odd-period cycles among the aperiodic orbits. Here is a stable 3-cycle at $a = 2.840$.

it is still possible to make short term predictions since the values of a few iterations with slightly differing initial conditions will remain reasonably close in the short term. Most people experience this phenomenon daily when they watch the weather forecast on TV, which is the forecaster's attempt to predict the behavior of an "apparently" chaotic dynamical system for the next few days. The word apparently is necessary at the present time as it has not yet been mathematically proven that the (Navier-Stokes) equations upon which weather forecasts are based, actually do produce chaotic solutions or not. But anyone who has had their day at the beach or picnic rained out will feel that the weather is indeed an example of a chaotic dynamical system.

By taking two points initially very close together in this fashion, it is possible to measure the rate of convergence or divergence of the ensuing trajectories. As it turns out, the rate of convergence or divergence increases in an exponential fashion, and this rate has a value equal to ae^{bn}, where $a > 0$ and b are constants and n is the number of the iteration. If b is negative, the trajectories of two neighboring points are converging at an exponential rate, and if b is positive, the values are diverging at an exponential rate. This divergence was illustrated in the preceding example, and an example of the case where the trajectories are converging exponentially is given by taking $a = 2.51$ and initial values $x_0 = 0.2$, $y_0 = 0.3$. Then after 100 iterations of each trajectory, we find that *both* are yielding the same cycle of four values to 12 decimal places:

$$
\begin{gathered}
0.528209070336 \\
1.153711733697 \\
0.708590766456 \\
1.226880395750
\end{gathered}
$$

This is simply the period 4 attractor we saw above in Fig. 2.4. Note, however, that by taking yet more decimal places there will be slight differences in the values of the iterates of the two trajectories, but these will gradually diminish as we take more and more iterations and the two trajectories converge.

The value b is called a *local Lyapunov exponent*: When $b < 0$, neighboring trajectories converge as in the preceding case, and when $b > 0$, trajectories diverge as in the case of the chaotic attractor.

Although a glance at the data shows that the trajectory values of the chaos example initially are steadily growing apart, there will be times further on in the iterations when values will become closer, move apart, come closer again, and so on. So what is needed is an average taken over the entire attractor and that is given by the *Lyapunov exponent.* It can be calculated by the formula

$$\lambda = \lim_{N \to \infty} \frac{1}{N} \sum_{n=0}^{N-1} \log_e \left| \frac{dx_{n+1}}{dx_n} \right| ,$$

where λ will be approximated by taking N reasonably large, so that in the case of our dynamical system we have

$$\lambda \approx \frac{1}{N} \sum_{n=0}^{N-1} \log_e |1 + a - 2ax_n| , \qquad \left(x_{n+1} = (1+a)x_n - ax_n^2\right) .$$

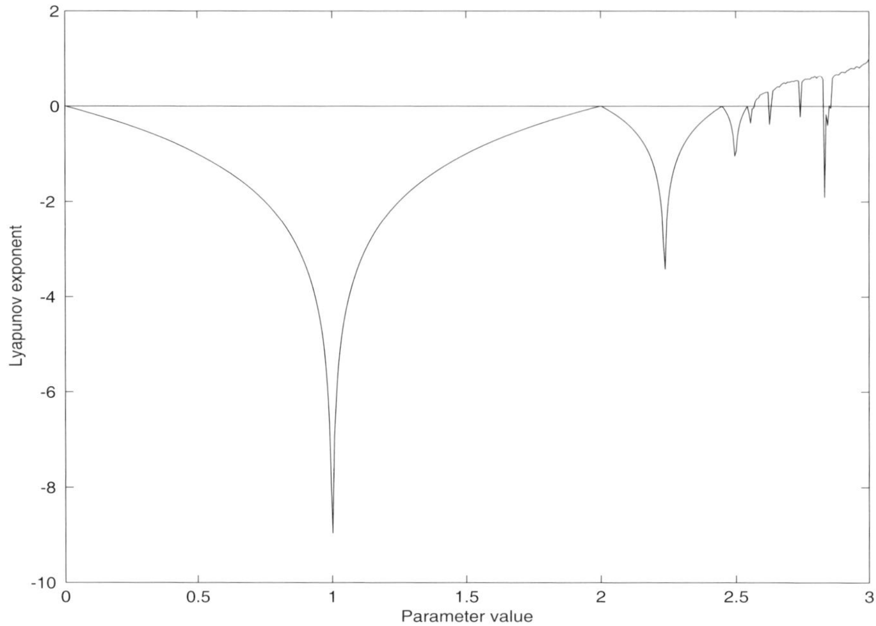

Figure 2.7 A graph of the value of the Lyapunov exponent λ for the parameter range $0 < a < 3$. When $\lambda < 0$ the dynamical system is stable, and for $\lambda > 0$ the system is chaotic. Note that there are intervals of stability in the chaotic regime beyond $a = 2.5699$.

This gives an average global measure of the convergent or divergent nature of the dynamical system. A nice account of the mathematical development leading up to the formulation of this equation can be found in Williams [2001]. From Fig. 2.7 one can observe the following behavior of our dynamical system:

$\lambda < 0$: The iterations of the system approach a stable fixed point or a stable periodic orbit.

$\lambda = 0$: This occurs in our system at bifuracation points, such as when $a = 2$.

$\lambda > 0$: The dynamical system is unstable and chaotic as we saw was the case in various regions of the parameter space after 2.5699 and when we took $a = 3$.

One important thing to note about chaos is that it is the result of a deterministic system. If you start with precisely the same initial value

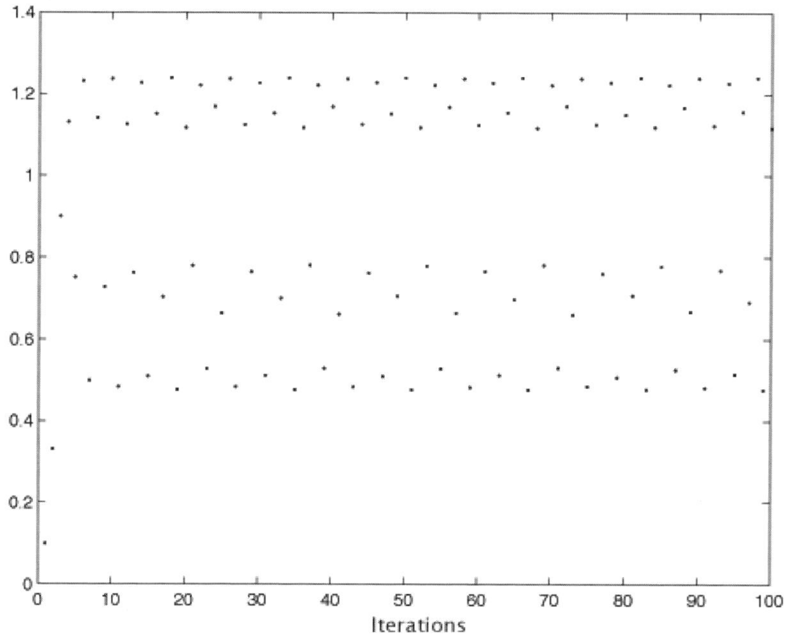

Figure 2.8 The strange attractor when the parameter value is set at $a = 2.5699$. The values of the orbit form a Cantor set.

x_0, then the resulting orbit, even though it lacks periodicity, will be precisely the same. However, any slight deviation, no matter how small, will eventually lead to vastly different values appearing in the orbit.

Another interesting phenomenon occurs at the starting point of the chaotic regime when $a = 2.5699...$. Here we obtain what is commonly known as a *strange attractor* in that the values of the orbit of the attractor form a Cantor set. Typically strange attractors have a fractal structure and may have chaotic dynamics, although some are fractal but nonchaotic. On the other hand, the terms "strange" and "chaotic" are often used synonymously (Williams [2001])

Magnifying the scale depicted in Fig. 2.8 a hundred-fold (and taking more iterations accordingly), we would see some of the "arms" of the attractor resolving into more sets of arms, and even further magnification would resolve some the resulting arms into yet further ones, and so forth. This type of infinite regression is typical of the Cantor set encountered in Section 1.5.

We are now ready to enter the remarkable world of cellular automata.

CHAPTER 3

ONE-DIMENSIONAL CELLULAR AUTOMATA

Cellular automata may be viewed as computers, in which data represented by initial configurations is processed by time evolution.
— *Stephen Wolfram* – Statistical Mechanics of Cellular Automata

The only reason for time is so that everything doesn't happen at once.
— *Albert Einstein*

3.1 THE CELLULAR AUTOMATON

We will consider a lattice network of cells that are most commonly square in shape, but the cells can be hexagonal and other shapes as well. Each cell can exist (not simultaneously) in k different states, where k is a finite number equal to or greater than 2. One of these states has a special status and will be known as the "quiescent state." The simplest case where each cell can exist in two possible states can be denoted by the symbols 0 and 1 and graphically by white and black, respectively. In more anthropomorphic terms, we can think of cells in the 0 (white/quiescent) state as "dead" and those in the 1 (black) state as "alive" (see Fig. 3.1).

The lattice of cells can be $n(\geq 1)$ dimensional, but most of the work on cellular automata (CA) has been for one and two dimensions, and we will have no recourse for any higher dimensions. In the sequel, we shall generally consider square cells as the basic unit of our automata. In the one-dimensional case, these form a row of adjacent boxes. In principle, the number of boxes in any array is infinite, but for practical purposes it will simply be taken sufficiently large to illustrate the behavior in question, often with certain conditions imposed on the cells along the boundaries. In some cases, intentional restrictions will be made on the size of the lattice as well.

In order for the cells of our lattice to evolve, we need time to change. This is done in an unusual manner by considering changes to the states of cells in the lattice only at discrete moments in time, that is, at *time steps* $t = 0, 1, 2, 3, ...$, as in the ticking of a clock. The time $t = 0$ is commonly reserved for the initial time period before any change of the cells' states has taken place. One further ingredient that is needed for our cellular lattice to evolve with discrete time steps is a local rule or *local transition function* governing how each cell alters its state from the present instant of time to the next based on a set of rules that take into account the cell's current state and the current state of its neighbors. Of course, which cells are considered to be neighbors needs to be precisely defined. This alteration of cell states takes place *synchronously* for all cells in the lattice. The transition function can be deterministic or probabilistic, but in most cases (some exceptions are models discussed in Chapter 5 on applications) we will consider only the former. The lattice of cells, the set of allowable states, together with the transition function is called a *cellular automaton*. Any assignment of state values to the entire lattice of cells by the transition function results in a *configuration* at any particular time step.

Therefore the three fundamental features of a cellular automaton are:

- *uniformity*: all cell states are updated by the same set of rules;
- *synchronicity*: all cell states are updated simultaneously;
- *locality*: the rules are local in nature.

Figure 3.1 A two-dimensional configuration with each cell taking one of two state values, black or white according to a particular local transition function. The first printed reference to a system such as the above, even using the word *automata*, is to be found in the paper by Ulam [1950].

There has been some recent work on the *asynchronous* updating of cell states (cf., e.g., Bersini and Detour [1994], Ingerson and Buvel [1984], Lumer and Nicolis [1994], Schönfisch and de Roos [1999]). This can be achieved in a number of different ways and is discussed in Section 4.4.

Other liberties with the rules of a cellular automaton can be taken. In one interesting example, galaxy formation has been simulated by Shulman and Seiden [1986] using a CA approach on a polar grid whose rings rotate at different rates (see Fig. 3.2). Theirs was a simple percolation model and achieves the basic structure of a spiral galaxy without a detailed analysis of the astrophysical dynamics involved.

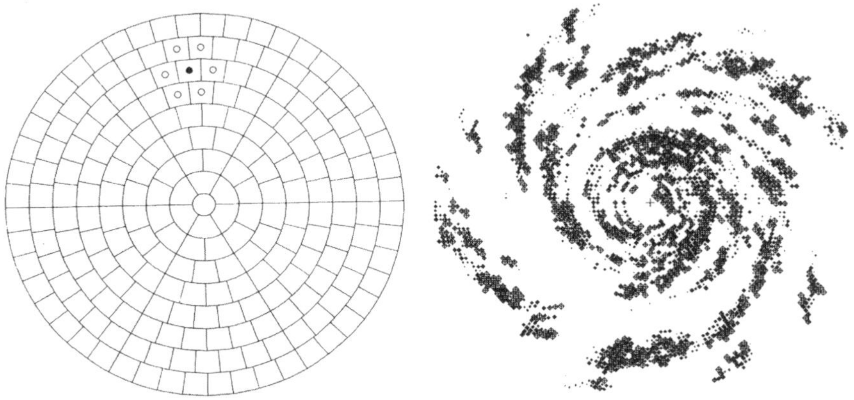

Figure 3.2 In this setting the neighbors of each cell change due to the differential rotation of the rings of the polar grid that is used to emulate galaxy formation. The black circle is an active region of star formation which induces star formation in its neighbors with a certain probability at the next time step. At right is a typical galaxy simulation.

In another digression from the conventional CA definition, Moshe Sipper [1994] considered different rules for different cells as well as the evolution of rules over time. There will be other instances in the sequel when we will stray slightly from strict adherence to the classical cellular automaton definition.

A fundamental precept of cellular automata is that the local transition function determining the state of each individual cell at a particular time step should be based upon the state of those cells in its immediate neighborhood at the previous time step or even previous time steps. Thus the rules are strictly local in nature and each cell becomes an information processing unit integrating the state of the cells around it and altering its own state in unison with all the others at the next time step in accordance with the stipulated rule. Thus, many global patterns of the system are an *emergent* feature of the effect of the locally defined transition function. That is, complex global features can emerge from the strictly local interaction of individual cells each of which is only aware of its immediate environment. This emergent behavior will be discussed more comprehensively in Chapter 6, but it is a salient feature that one should always be aware of in our development of cellular automata theory.

Figure 3.3 The eight possible neighborhood-states with $r = 1$ and $k = 2$. Here, as will be our usual convention, black is 1 and white is 0.

·	·	·	$c_{i-1}(t)$	$c_i(t)$	$c_{i+1}(t)$	·	·	·

Figure 3.4 The cell states of the central cell c_i and its two nearest neighbors c_{i-1} and c_{i+1} at the time step t.

3.2 TRANSITION FUNCTIONS

Most of the dynamical features of cellular automata can be found in the study of the one-dimensional case. Here we define a neighborhood of a cell c having radius (range) r as the r cells to the left of c and the same number of cells to the right of c. Counting c itself, this neighborhood contains $2r + 1$ cells. In the simplest case $r = 1$ and $k = 2$ for the allowable states. In this instance, a three-cell neighborhood with two different states 0 and 1 for each cell can be expressed in $2^3 = 8$ different ways. All eight neighborhood-states with $r = 1$ and $k = 2$ are illustrated below (Fig. 3.3), and in general there are k^{2r+1} one-dimensional neighborhood-states.

Let us adopt this notation: $c_i(t)$ denotes the state of the ith cell at time t (Fig. 3.4).

At the next time step, $t+1$, the cell state will be $c_i(t+1)$. Mathematically we can express the dependence of a cell's state at time step $t+1$ on the state of its left-hand and right-hand nearest neighbors $c_{i-1}(t)$ and $c_{i+1}(t)$ at time step t by the relation

$$c_i(t+1) = \varphi\left[c_{i-1}(t), c_i(t), c_{i+1}(t)\right],$$

where φ is the local transition function. For example, consider the simple transition rule

$$c_i(t+1) = c_{i-1}(t) + c_i(t) + c_{i+1}(t) \mod 2, \tag{3.1}$$

where $\mathrm{mod}\, 2$ means taking the remainder after division of the indicated sum by 2, resulting in either 0 or 1. We can put this rule into a *transition table* format by adding $c_{i-1}(t) + c_i(t) + c_{i+1}(t) \mod 2$ for the eight possible different input values:

$c_{i-1}(t)$	$c_i(t)$	$c_{i+1}(t)$	$c_i(t+1)$
1	1	1	1
1	1	0	0
1	0	1	0
1	0	0	1
0	1	1	0
0	1	0	1
0	0	1	1
0	0	0	0

(3.2)

A very convenient way to illustrate the allowable rules for one-dimensional cellular automata with $r = 1$ and $k = 2$ is to indicate the state (color) of the middle cell at the next time step, given the state (color) of itself and its two nearest neighbors (Fig. 3.5).

Here the middle cell at time t (in the top row) has its state altered according to the state of its two nearest neighbors and its own state to yield the cell's new state at time $t + 1$ (bottom row). This also represents the rule in Equation 3.1. Observe that in four of the eight cases a black cell appears. One consequence of this is that if we took a disordered array of a large number of cell sites, then the average fraction (or *density*) of black cell sites that evolve after one iteration will be 0.5.

The cellular automaton defined by this rule has the graphical representation depicted in Fig. 3.6 starting with a single black cell with each subsequent generation of the automaton appearing on the next line down.

So how many possible rules are there for $r = 1$ and $k = 2$? Since there are eight possible neighborhood-states of three cells and each of these results in two possible state outcomes for the middle cell, there are $2^8 = 256$ possible transition rules by which this can be achieved. Wolfram described these as *elementary* cellular automata. We can also identify the rule given in the form of the transition table 3.2, by its eight output states (or *rulestring*) 10010110, which in base 2 happens to be the number 150. So this rule is referred to as Rule 150 and is the rule

Figure 3.5

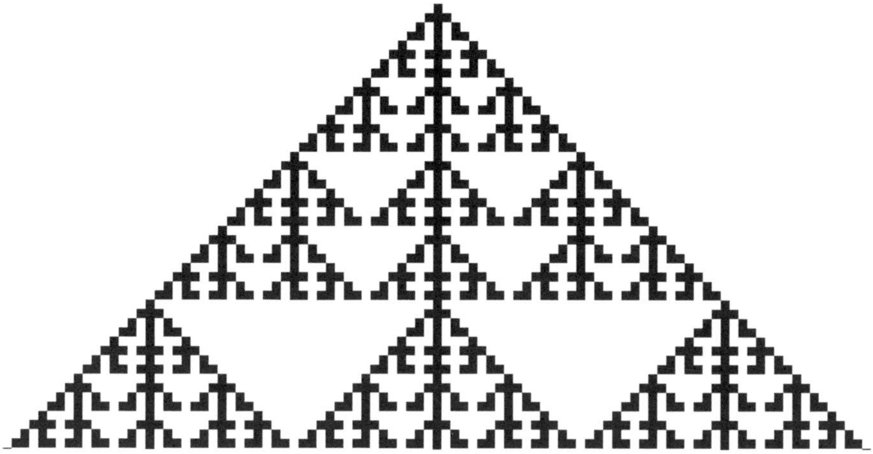

Figure 3.6 By considering a white cell as being in state 0 and a black cell in state 1, we can depict the evolution of the rule in the text starting with a single black cell. Each new line downward represents the evolution of the automaton at the next time step.

above depicted now in five different formats: Equation 3.1, transition table 3.2, the rulestring, and Figs. 3.5 and 3.6.

In a similar fashion, all 256 elementary cellular automata can be numbered starting from Figs. 3.7 to 3.10.

A complete illustration of all 256 elementary cellular automata starting with a standard initial condition of one black cell is given in Appendix A. Many of the salient features found in cellular automata theory can be observed in these elementary ones and we shall refer to them often. These elementary cellular automata are examples of *first order automata* in the sense that the state of a cell at time step $t+1$ only depends on the state of its neighbors at the previous time step t. Whereas in a *second order automaton* a cell's state at time step $t+1$ is rather more demanding and depends on the state of its neighbors at time steps $t-1$ as well as t, analogous to the way the Fibonacci sequence was formed. Unless otherwise specified, we will mainly be considering first order automata, although second order automata will come into play in our discussion of reversibility.

Perhaps you are thinking that this is not so much to work with, so if we merely increase the radius to $r=2$, then there are $2^5 = 32$ possible neighborhood-states and $2^{32} = 4,294,967,296$ possible rules. On the

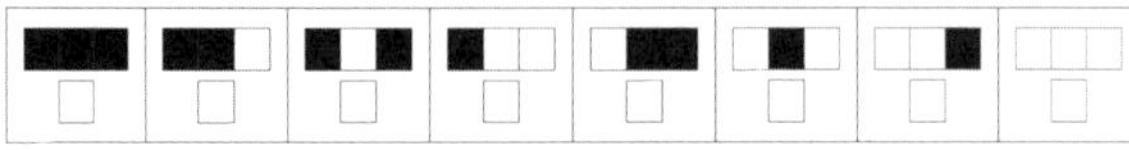

= Rule 0

Figure 3.7

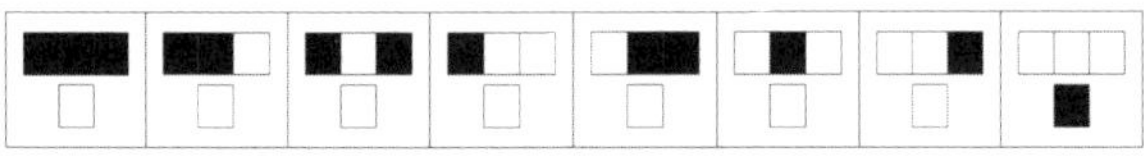

= Rule 1

Figure 3.8

⋮

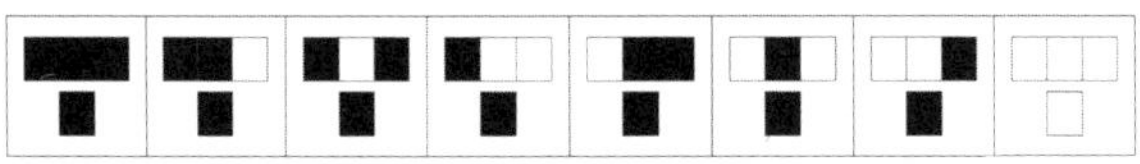

= Rule 254

Figure 3.9

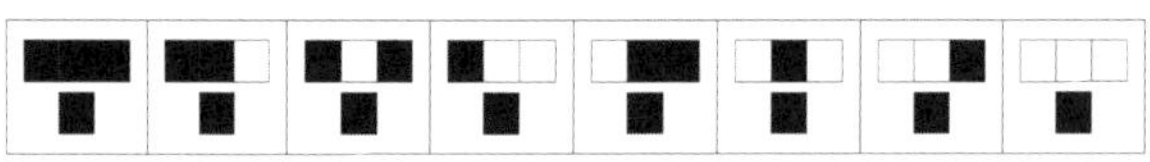

= Rule 255

Figure 3.10

other hand, if we keep $r = 1$ and merely increase the allowable states per cell to 3, then there are $3^3 = 27$ possible neighborhood-states and an astronomical $3^{27} = 7,625,597,484,987$ possible rules! That should be plenty.

3.3 TOTALISTIC RULES

This class of rules is defined by taking the local transition function as some function of the *sum* of the values of the neighborhood cell sites, where the neighborhood includes the central cell and hence its value is included in the sum. This definition applies to higher dimensions as well. So, for example, if $r = 1$, $k = 2$, and denoting φ as the local transition function, we have

$$c_i(t+1) = \varphi\left[c_{i-1}(t) + c_i(t) + c_{i+1}(t)\right].$$

An example is the Rule 150 above given by

$$c_i(t+1) = c_{i-1}(t) + c_i(t) + c_{i+1}(t) \mod 2.$$

So how many such totalistic rules are there with $r = 1$, $k = 2$? With the two states 0 and 1 for each of the three cells, there are just the four possible totals of the cell states: $0, 1, 2$, and 3. For example, our Rule 150 can also be depicted by the following transition table:

sum (t)	3	2	1	0
$c_i(t+1)$	1	0	1	0

So we can see that in general each of the four sums accruing at time step t can yield at the next time step either a 0 or 1, thus giving $2^4 = 16$ possible totalistic rules, of which the preceding Rule 150 is one.

Breaking away from the elementary cellular automata and allowing three states per cell ($k = 3$) but maintaining nearest neighbors ($r = 1$), let us take cell states $0 =$ white, $1 =$ gray, and $2 =$ black. Thus there are seven possible sums: $0, 1, 2, 3, 4, 5$, and 6, each of which can yield a 0,1, or 2 at the next time step. An example is:

sum (t)	6	5	4	3	2	1	0
$c_i(t+1)$	1	1	0	1	2	1	0

and in fact, there are $3^7 = 2187$ such totalistic rules. In order to systematically enumerate them, the cellular automaton number is the base 3 value of $c_i(t+1)$ in the second row as in the preceding table. The number given above, 1101210, is the value 1020 in base 3 and so this automaton is labeled Code 1020 and has the appearance in Fig. 3.11 starting with a single gray cell.

There are also what are called *outer totalistic* rules in which the value of the central cell at the next time step depends on the current value of the central cell as well as just the sum of the values of the other neighbors. The Game of Life discussed in the next chapter is one such well-known example.

3.4 BOUNDARY CONDITIONS

There is the small matter of what should be the neighborhood of the cells at the extreme left and right of the lattice. In many cases we will take the lattice to be sufficiently large in extent so that these cells do not come into our considerations and the lattice can be considered to be effectively infinite. However, in some cases the lattice will be finite in extent and there are three main types of boundary conditions:

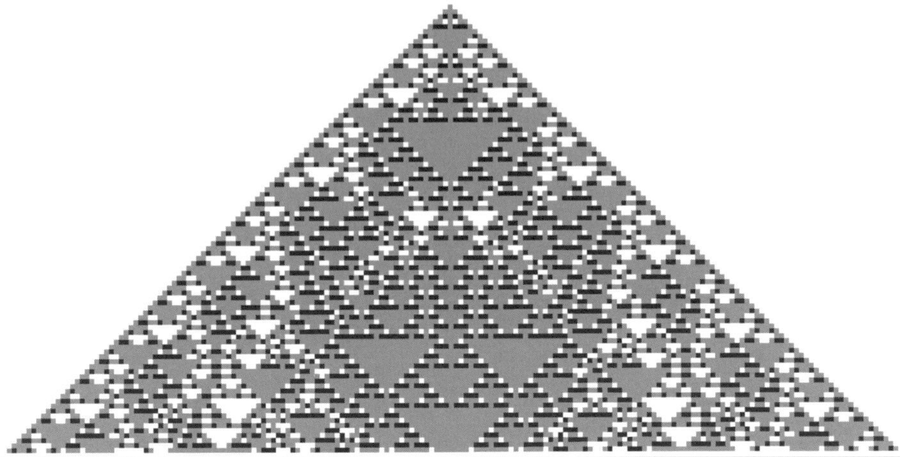

Figure 3.11 Totalistic Code 1020 = 1101210 in base 3, generated from a single gray cell (cell state = 1) with three states per cell and nearest neighbors.

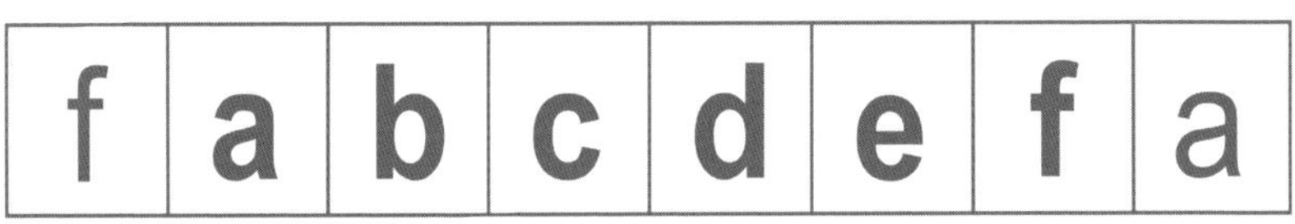

Figure 3.12 For the finite one-dimensional array of cells labeled a, b, c, d, e, f, periodic boundary conditions ensure that cell a has the left-hand neighbor f and cell f has the right-hand neighbor a as if the array were joined into a loop.

(i) *Periodic* (also known as "wrap"): Here we consider the two cells at the extreme left and right ends to be neighbors of each other (Fig. 3.12), thus making the lattice into a continuous ring in the one-dimensional case. As well, it is also possible to prescribe the array in the vertical direction (in the two-dimensional case) so that the top and bottom cells are also neighbors, thus making the cellular array into the shape of a torus, or donut. Periodic boundary conditions are the best at simulating an infinite array since no boundary is apparent to any cell.

(ii) *Reflective*: In this case the cells' states at the extremities are repeated in order to provide neighboring cell values for them (Fig. 3.13).

(iii) *Fixed*: In some instances, we want the boundary values to take on some fixed value such as when modelling a problem in heat conduction

a	a	b	c	d	e	f	f

Figure 3.13 With reflective boundary conditions, the same array of cells will have the first and last cells repeated in order to provide neighbors for them.

(see Section 5.5.6). Thus the values along the boundary remain constant throughout the simulation.

3.5 SOME ELEMENTARY CELLULAR AUTOMATA

In the sequel we will consider various elementary cellular automata that have an initial condition of a single black (i.e., state 1) cell, all the others being white (state 0). Of course we will not discuss all 256 cases, nor is this necessary, as for example, Rule 0 (Fig. 3.7) only produces cells that are in state 0 and is therefore of zero interest.

Rule 1 (Fig. 3.8) has the distinguishing feature that a neighborhood of cells entirely in state 0 (quiescent) spawn a live cell at the next time step. In some theoretical considerations (see Section 4.1.6) this condition will not be permitted. This yields the automaton depicted in Fig. 3.14, starting from a single black cell.

In principle, this array must be infinite in extent to the left and right of the starting cell.

Let us skip now to look at Rule 30:

$$c_i(t+1) = [c_{i-1}(t) + c_i(t) + c_{i+1}(t) + c_i(t)c_{i+1}(t)] \mod 2,$$

where we are immediately struck by its quite random appearance (Fig. 3.15).

Considering the central column of cells, which have states 1101110011..., Wolfram [1986] applied various statistical tests and found the sequence of 1's and 0's to be completely random. Of course this is the deterministic randomness discussed earlier, and every time Rule 30 is executed it will produce the same sequence of digits along the central column. Nevertheless, no pattern whatsoever can be discerned

Figure 3.14 Rule 1 starting with a single black cell illustrating its repetitive nature.

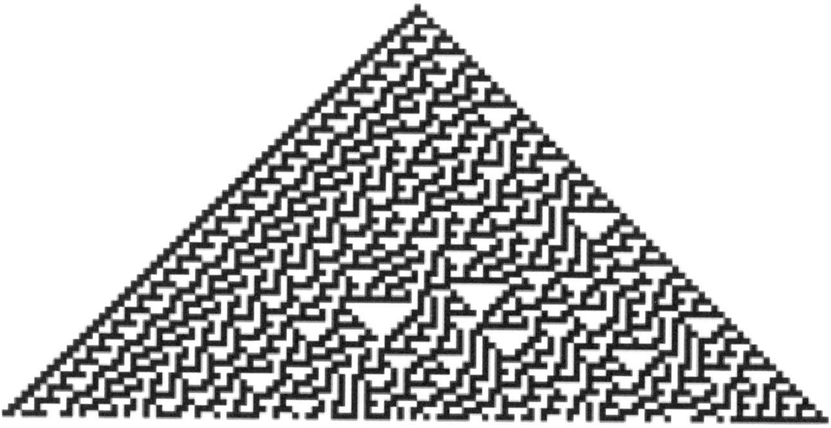

Figure 3.15 Rule 30 depicting a rather random appearance containing triangles of various sizes.

in these digits in the sense mentioned, and so they are considered random.

To consider the randomness of the central column of (in this example taking a sample of 960) cells a bit further, if it is truly random, then one would expect that the number of 0's and 1's in any long sequence to be roughly equal, so that each digit occurs roughly half the time. Likewise, we should find approximately equal numbers of the 4 pairs of digits 00, 01, 10, 11 (among the 480 possible pairs of digits), so that the frequency of occurrence (called the *block frequency*) of each pair is $\frac{1}{4}$, and for the 8 triples of digits 000, 001, ..., 111 (among the 320 triples) each should have a block frequency approximately of $\frac{1}{8}$, and similarly for the 16 quadruples 0000, 0001, ..., 1111 (among the 240 quadruples), each should have a block frequency of $\frac{1}{16}$. To test this, a sequence of 960 cells in length from the central column was sampled and the block frequency of each block length up to length 4 was calculated. Here is a table of the counts for each block size:

Block size	**Observed count**							
1	498	462						
2	134	116	114	116				
3	55	38	34	41	41	38	28	45
4	22	16	16	19	9	16	16	15
	19	11	11	13	17	19		

The *expected* counts E_i for each block size respectively are based on their probability of occurrence p_i, with $E_i = p_i N$, and $N = 960$:

Block size	**Expected count**
1	$E_1 = E_2 = 1/2 \times 960 = 480;$
2	$E_1 = E_2 = E_3 = E_4 = 1/4 \times 480 = 120;$
3	$E_1 = E_2 = \cdots = E_8 = 1/8 \times 320 = 40;$
4	$E_1 = E_2 = \cdots = E_{16} = 1/16 \times 240 = 15.$

In order to compare the observed counts for each particular block size with the expected counts in each case, we use a statistical tool known as the *chi-squared test.* This is just a general measure of how much the observed values deviate from the expected ones determined by their probabilities and is given by

$$\chi^2 = \sum_{i=1}^{k} \frac{(O_i - E_i)^2}{E_i},$$

where O_i represents the *observed* counts for a particular block size, E_i are the expected counts, and k is the number of different blocks considered in each case. The formula gives the value denoted by χ^2, called *chi squared*. So, for example, if we consider blocks of size 3, then $k = 8$ and each E_i equals 40. Based on the data above we find:

Block size 1: $\chi^2 = 1.35$

Block size 2: $\chi^2 = 2.2$

Block size 3: $\chi^2 = 11$

Block size 4: $\chi^2 = 16$

Next we need to interpret these numbers to see if they mean that our data are actually random. In order to do this, there are standard tables for comparing your chi-squared values with truly random ones. So, for example, if we look up the value of $\chi^2 = 1.35$ for our block size 1, we find that between 20 and 30% of the time a truly random sequence will have an even larger chi-squared value. Since the χ^2 value is a measure of the deviation from the theoretical expected values, this is indeed a good indication of randomness for Rule 30. Taking, say, block size 4 and our value of $\chi^2 = 16$, we find from the tables that between 30 and 50% of the time a random sequence will have a larger chi-squared value, again a strong indication of randomness for Rule 30. The other χ^2 values give similar results and we may confidently conclude that the pattern of black and white cells along the central column of Rule 30 is indeed random.

Various other tests for randomness are discussed in the book by Donald Knuth [1981], and again Rule 30 passes all of these tests as well (Wolfram [1986 and 1994]).

Rule 30 has also been proposed by Wolfram as the basis for an encryption system. To implement such a system the first step is to transcribe all of the plaintext message into 8-bit ASCII characters. For example, the letter "a" is denoted by 01100001 ($= 97$), "b" is 01100010 ($= 98$), and "space" is 00100000 ($= 32$), and so forth. For ease of handling purposes we put this into an $8 \times n$ array, where n is the length of the message (including spaces and punctuation). Thus the plaintext message

Once upon a time...

is put into an array where each row represents a letter with white = 0, black = 1; see Fig. 3.16.

So the first letter O, which is $79 = 01001111$, appears on the first row and on the fifth row we have a space, which is $32 = 00100000$,

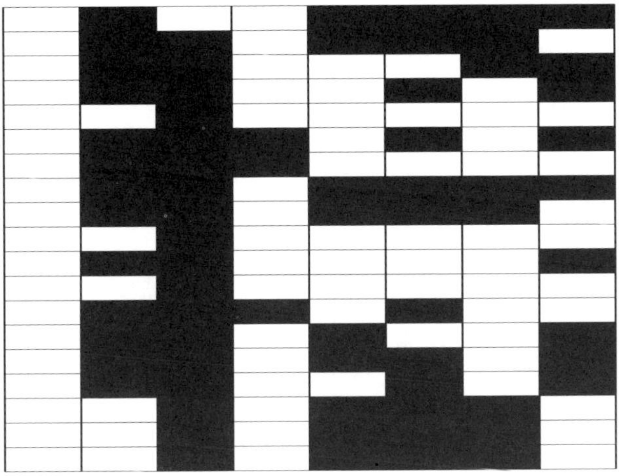

Figure 3.16 The plaintext message "Once upon a time..." transcribed into 8-bit ASCII with 0 = white and 1 = black.

and so forth. Let us denote an abitrary member of this array by a_i. To encrypt this message we take some specific set of random initial values and evolve Rule 30 from them. Now take the cell values of 1's and 0's of the central (or any other) column and denote these values by b_i. (Actually taking every other value from a particular column provides a higher level of security.) In Fig. 3.17 are the values b_i of Rule 30 derived from the initial condition 1772712395 expressed as a 31-digit binary number taking every second cell value of the 16th column that we will use to encrypt our message.

Then we simply apply the EXCLUSIVE OR (XOR) operation

$$a_i \vee^* b_i = c_i$$

taking elements from the plaintext array and Rule 30 array consecutively, with 1 = True, 0 = False. This results in the ciphertext c_i, which we also put into an array format (Fig. 3.18).

For example, starting with the top row and taking consecutively the elements of the message and Rule 30, forming $a_i \vee^* b_i = c_i$, we have $0 \vee^* 0 = 0$, $1 \vee^* 1 = 0$, $0 \vee^* 1 = 1$, and so on. The XOR is slightly more restrictive than the usual $\vee$ operation and differs from it only in so far as, if P and Q are both true, then $P \vee^* Q$ is now false. If either just one of P or Q is true, then as for the usual disjunction $P \vee^* Q$ is likewise true.

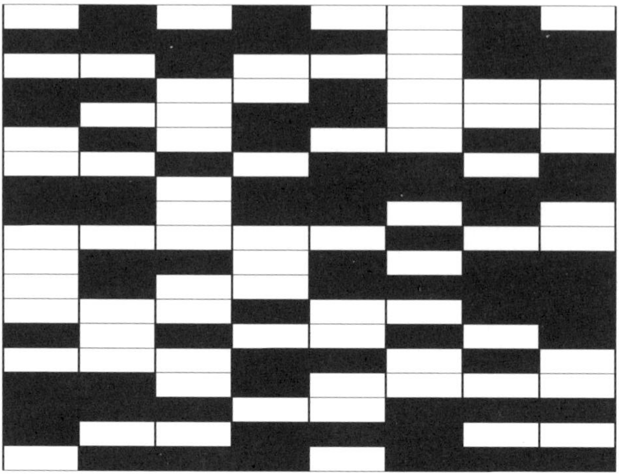

Figure 3.17 The values given by Rule 30 derived from the initial condition 1772712395 expressed as a binary digit number and taking every second number of the 16th column.

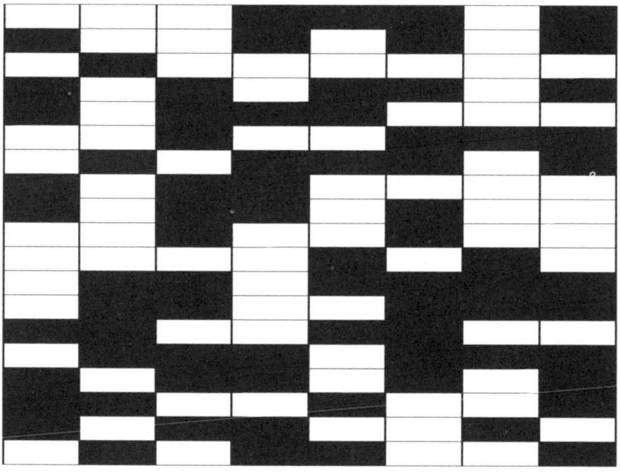

Figure 3.18 The ciphertext of the original message encoded using Rule 30 according to the algorithm explained in the text.

The ciphertext message can now be transmitted. In order to decode the message, one simply has to run Rule 30 taking the *same set* of random initial conditions and applying the XOR operation consecutively to the ciphertext and the same elements b_i of Rule 30 that were used for the encryption:

$$c_i \vee^* b_i = a_i.$$

The result of this second application of the XOR operation returns the original values a_i, which in this case translates into *Once upon a time...* .

The reason this works is that two consecutive applications of the XOR operation returns the original value. For example, for a value of 1 in the text and a value of 1 from Rule 30, we have $1 \vee^* 1 = 0$, and taking this result again with 1 from Rule 30, we have $0 \vee^* 1 = 1$, which is the original text value. The other three possibilities are verified similarly.

One disadvantage to this method of encipherment is that the random initial conditions for Rule 30 must be communicated to the message recipient in advance. This has been a weakness of secret codes over the centuries that have relied on the "key" having to be known to both parties to encipher and decipher a message. There is always the danger that the key will be discovered by those for whom the message is not intended. Sometimes this key would change daily and so-called *one-time pads* would be used for this purpose. New developments in quantum cryptography are making the distribution of the key absolutely secure (see Bennett, Brassard, and Ekert [1992]). Individual photons are sent by fiber optic cable between the two parties and any attempt to eavesdrop on these photons will perturb them in such a way that can be detected by the intended recipients.

Another form of encryption, *public key encryption*, based on the extreme difficulty of factoring the product of two large prime numbers into its constituent parts, is considered uncrackable at the present time. This method was discovered by Clifford Cocks of the British Government Communication Headquarters in 1973 and independently in 1977 by Ron **R**ivest, Adi **S**hamir, and Leonard **A**dleman of MIT and has become known as RSA. In the quaint jargon of cryptography, Alice creates a public encryption key which is just the product of two prime numbers, say $N = p \times q = 10151 \times 14533 = 147524483$, but keeping the values of p and q secret. This allows anyone who knows the key to send Alice a message. For instance, if Bob wants to send Alice a message, he simply uses a particular mathematical algorithm (called a *one-way function*, also publically known) that invokes Alice's key $N = 147524483$ to encrypt the plaintext and send it to her. To decipher the message requires knowing the values of p and q so that Alice is the only one who can do this. The ciphertext message is secure

Figure 3.19 Rule 90 after 64 steps of its evolution.

because, in practice, the prime numbers that Alice has chosen, p and q, are very large (say 100 digits long!), and once multiplied together, the number N cannot be factored into its constituent parts by even the most powerful computers. It is conceivable that someday there will be found some mathematical algorithm for achieving the factorization of extremely large numbers in a reasonable period of time, but until then the RSA encryption method is totally secure, although encrypting large amounts of data with it is cumbersome. It is the *defacto* standard for secure transactions on the Internet. Some attempts have been made to base public key encryption on cellular automata (see Tim Tyler's website: `http://cell-auto.com/`). On the other hand, a CA scheme for the authentication of digital messages and images has been developed by Mukherjee *et al.* [2002]. A modern day fascinating account of the history of cryptography is Simon Singh's *The Code Book* [1999].

Several elementary cellular automata (Rules 18, 22, 26, 82, 146, 154, 210, and 218 among others, see Appendix A) have the nested triangle appearance of Rule 90, which is given by the formula $c_i(t+1) = c_{i-1}(t) + c_{i+1}(t) \mod 2$ (see Fig. 3.19).

In the limiting case, this automaton depicts the self-similar Sierpiński pattern and has fractal dimension 1.585. It has been shown by D. A. Lind [1984] that starting from any initial configuration the limiting frequency of 1's and 0's is always $\frac{1}{2}$ for each. This is an exceptional result because in general it is impossible to say much of anything about the long term evolution of a particular cellular automaton.

Rule 90 also has a connection with Pascal's triangle, a familiar object in elementary mathematics. A consequence of the well-known binomial theorem is the expansion

$$(1+x)^n = 1+nx+\frac{n(n-1)}{1\cdot 2}x^2+\frac{n(n-1)(n-2)}{1\cdot 2\cdot 3}x^3+\cdots+nx^{n-1}+x^n$$

for positive integer values of n. For example,

$$(1+x)^5 = 1+5x+10x^2+10x^3+5x^4+x^5.$$

Pascal's triangle derives the coefficients of the expansion by starting with a 1 at the apex of the triangle, then the first binomial expansion of $(1+x)^1$ has coefficients $(1,1)$ which form the next row of the triangle and the coefficients of $(1+x)^2$ which are $(1,2,1)$ are on next row, where the 2 is the sum of the previous two 1's. The next row will have a 1 at either end and between them we add the first 1 and 2 from the previous row to get 3 as the next digit, then add the 2 and 1 from the previous row to get another 3, so that the third row is $(1,3,3,1)$. In general, we always start and finish a row with a 1 and the digits in between are obtained by adding the consecutive digits of the previous row. This generates a triangle of digits known as Pascal's triangle that gives the binomial coefficients of $(1+x)^n$. If we overlay this triangle of digits with a grid of square cells and color the cells black that have odd numbers in them, then we obtain the pattern of Rule 90.

Rule 110 is quite unremarkable in appearance starting from a single black cell (Fig. 3.5).

Its rule structure (Fig. 3.21) gives nothing away regarding its remarkable potential, which Wolfram suspected would be nothing less than universal computation. And, indeed, during the 1990s his research assistant, Matthew Cook, demonstrated that Rule 110 was capable of universal computation! This was demonstrated by showing that Rule 110 could in fact emulate any of what are called *cyclic tag systems.* Since some of these can emulate any Turing machine and since some Turing machines are capable of universal computation, it logically follows that Rule 110 is likewise capable of universal computation (see Cook [2004]). This is truly a landmark result. Some complex interactions of Rule 110 can be seen in Fig. 3.22.

Rule 150 has been introduced in Section 3.2 and starting from a single black cell has the appearance depicted in Fig. 3.6. Remarkably, in the

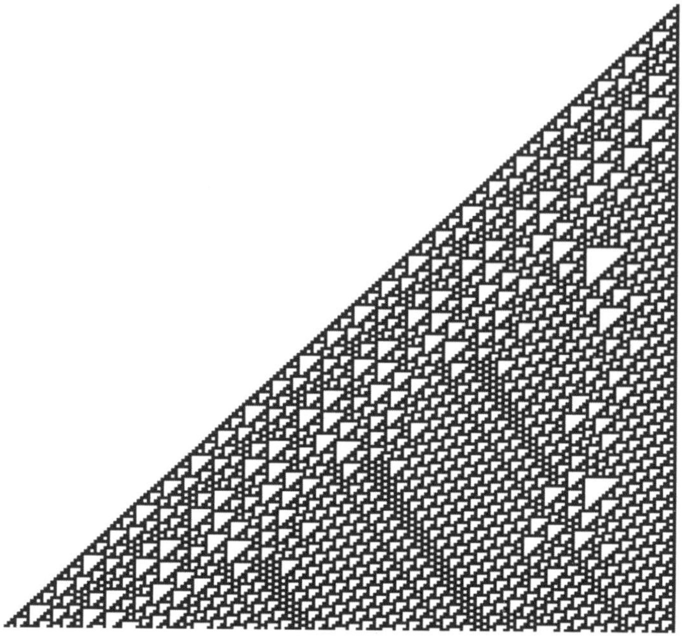

Figure 3.20 The evolution of Rule 110 starting with one black cell.

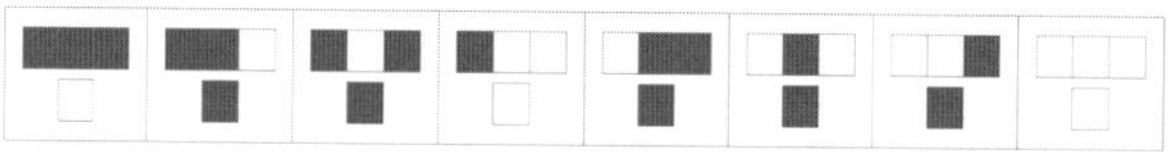

Figure 3.21 The inconspicuous looking Rule 110.

same way that Rule 90 is related to the coefficients of the binomial expansion, with odd-numbered cell entries being black, Rule 150 is related to the coefficients of the trinomial expansion

$$(1 + x + x^2)^n.$$

As an example, taking $n = 8$ gives the coefficients **1**, 8, 36, 112, 266, 504, 784, 1016, **1107**, 1016, 784, 504, 266, 112, 36, 8, **1** with the first, middle and last coefficients being the only odd ones. These determine which cells are black in the ninth row of Rule 150 and all other black cells are obtained in a similar fashion.

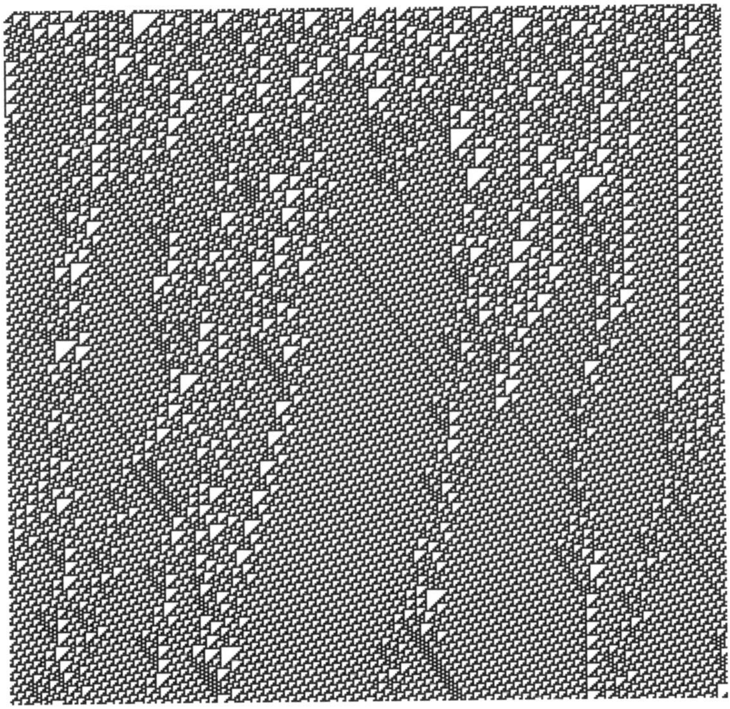

Figure 3.22 This is Rule 110 starting with random initial conditions. Here we begin to sense some of the complex interactions that the rule is capable of.

3.6 ADDITIVITY

Some rules like Rule 90 or Rule 150 have a special property of being what is called additive. Taking another look at Rule 90,

$$c_i(t+1) = c_{i-1}(t) + c_{i+1}(t) \mod 2,$$

we find that the value of the central cell at the next time step depends on a sum of the neighboring cells' states. We can extend the sum notion slightly to what mathematicians call a *linear combination* which for $k = 2$ and $r = 1$ is a sum of integer multiples of the neighborhood states,

$$c_i(t+1) = \alpha c_{i-1}(t) + \beta c_i(t) + \gamma c_{i+1}(t) \mod 2,$$

and the α, β, γ are integer constants 0 or 1. In general, for k states per cell there are 0 to $k-1$ values for the constants and all the arithmetic is

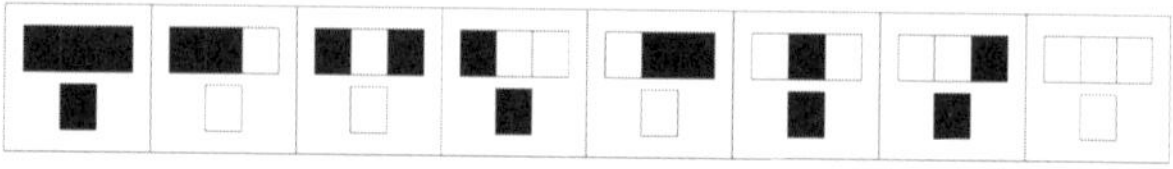

Figure 3.23 Transition function for Rule 150 displaying its additivity property.

taken $\mod k$. For k states and range r there are k^{2r+1} additive rules or $2^{2\cdot 1+1} = 8$ additive rules among the 256 elementary cellular automata. These are found by taking all possible choices of 0 and 1 for α, β, γ giving: Rule 0 $(\alpha = \beta = \gamma = 0)$, Rule 60 $(c_{i-1} + c_i)$, Rule 90 $(c_{i-1} + c_{i+1})$, Rule 102 $(c_i + c_{i+1})$, Rule 150 $(c_{i-1} + c_i + c_{i+1})$, Rule 170 (c_{i+1}), Rule 204 (c_i), Rule 240 (c_{i-1}).

The above notion of additivity can easily be seen by taking Rule 150, for example, and looking at its transition function (Fig. 3.23).

It is clear from the rule that the right-hand neighbor, central cell, and left-hand neighbor state values can be added $(\mod 2)$ to give the state value of the central cell at the next time step.

The additive CA all exhibit nesting behavior as illustrated by Rules 90 and 150 and also have the advantage that an algebraic analysis can be used to determine certain of their global properties (Martin, Odlyzko, and Wolfram [1984]). For a comprehensive overview of the subject together with a wide variety of CA applications, see the book by Chaudhuri *et al.* [1997].

3.7 REVERSIBILITY

It is thought that the laws of physics are time-reversible, in other words, if the state of a system is known at a given instant of time, it is possible to completely describe not only the future evolution of the system, but its past as well. This is due to the fact that in the equations describing these systems the time variable t can be replaced by $-t$ and the equations still remain valid. According to Einstein, "the distinction between past, present, and future is only an illusion, however persistent" (excerpt from a letter to the widow of his friend Michele Besso). This is seemingly at odds with the Second Law of Thermodynamics which describes irreversible processes that proceed with increasing entropy along an arrow of time. This dichotomy — the microscopic laws of physics are reversible, yet the macroscopic world is filled with irreversible phenomena — has exercised the minds of physicists and philosophers at least since the 1860s amid much controversy and confusion. To demonstrate the flavor of the confusion and controversy, let me just quote Jean Bricmont [1997], professor of theoretical physics at the Université catholique de Louvain:

> It is perfectly possible to give a natural account of irreversible phenomena on the basis of reversible fundamental laws, and of suitable assumptions about initial conditions. This was essentially done a century ago by Boltzmann, and despite numerous misunderstandings and misguided objections (some of them coming from famous scientists, such as Zermelo or Poincaré), his explanation still holds today. Yet [Ilya] Prigogine writes... "He (Boltzmann) was forced to conclude that the irreversibility postulated by thermodynamics was incompatible with the reversible laws of dynamics." This is in rather sharp contrast with Boltzmann's own words: "From the fact that the differential equations of mechanics are left unchanged by reversing the sign of time without anything else, Herr Ostwald concludes that the mechanical view of the world cannot explain why natural processes run preferentially in a definite direction. But such a view appears to me to overlook that mechanical events are determined not only by differential equations, but also by initial conditions...".

As Bricmont further elaborates, "irreversibility does not lead to a *contradiction* with the basic physical laws. Indeed, the laws of physics are always of the form given some initial conditions, here is the result after some time. But they never tell us how the world *is* or *evolves*. In order to account for that, one always needs to assume something about the initial conditions." Incidently, Jean Bricmont is also co-author of *Intellectual Impostures* (published in the United States as *Fashionable Nonsense*), an exposé of pseudo-intellectual, pseudo-scientific discourse that passes for profound philosophical thought.

In fact, there are some reversible cellular automata that do proceed from a disordered state to an ordered one, that is, from a state of high entropy to a state of low entropy. And of course the key here is to precisely know the correct initial conditions. In Fig. 3.24, the automaton proceeds in the first column from a state of randomness to a highly ordered state. How was this highly unusual state of affairs achieved? Namely, by running the inverse of the automaton (starting at the top of column 5) until a final state is reached and capturing the values of this final state for the initial conditions we started with in column 1.

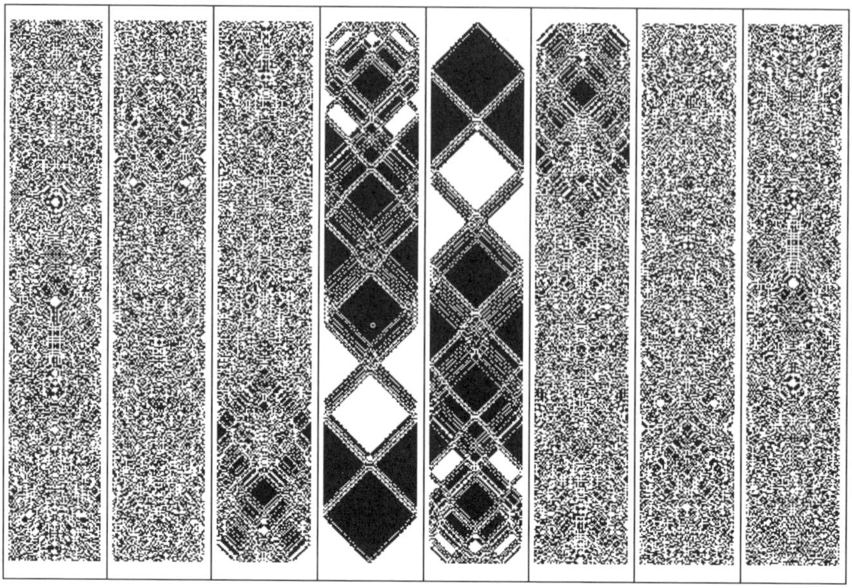

Figure 3.24 This reversible automaton proceeds from a completely disordered state to a highly ordered one (columns 1–4), contrary to the Second Law of Thermodynamics. This was achievable by knowing precisely the initial conditions to start with. So the trick is to run the inverse of the rule until the automaton becomes completely disordered (columns 5–8) and the final state of the system is used for the initial state of the original rule.

Thus in order to attain reversibility, it all depends on an exact knowledge of the initial conditions of the system. But therein lies the rub when it comes to actual physical systems. As Stephen Wolfram states [2002], "no reasonable experiment can ever involve setting up the kind of initial conditions that will lead to decreases in randomness, and that therefore all practical experiments will tend to show only increases in randomness... . It is this basic argument that I believe explains the observed validity of the Second Law of Thermodynamics."

By way of contrast, *irreversible* cellular automata may exhibit a decrease in the number of possible configurations over time (*compression of states*) and hence a decrease in entropy. This means that the number of reachable configurations diminishes over time and the system tends toward a more ordered state.

To address this point, let us consider a finite cellular automaton with N cells and two possible states per cell, so that there are 2^N possible configurations. If we consider our automaton to be initially completely disordered, then any configuration is equally likely, having a probability of occurrence of $1/2^N$. Thus the automation is in a state of maximum entropy. Now at the first time step let each one of the 2^N possible initial configurations evolve according to the transition rule and let us record each of the configurations thusly reached. Some of these configurations may now be the same, say one configuration appears m times among these 2^N configurations. Then its probability of occurrence at this time step becomes $m/2^N$. At each subsequent time step, we will compute the probability of occurrence in this fashion for each of the descendent configurations generated by the time evolution of the initial ensemble of configurations. Whenever some configurations appear multiple times, others will not appear at all and the probability of occurrence of these latter will be zero. We have illustrated a simplified version of this situation in the table below with the label of *all* the possible initial configurations of an automaton along the top row and time being vertical. So a is just the name for a configuration of black and white cells, likewise b, and so on. If our automaton was just 10 cells wide, there would indeed be $2^{10} = 1024$ such configurations, but the eight listed here will suffice to illustrate our point. Now the local transition function will give rise to the transformation of one configuration into another, say: $a \longrightarrow b$, $b \longrightarrow c$, $c \longrightarrow d$, $d \longrightarrow e$, $e \longrightarrow f$, $f \longrightarrow g$, $g \longrightarrow h$, $h \longrightarrow g$. This automaton is irreversible as configuration g has two predecessors, namely f and h.

	a	b	c	d	e	f	g	h
1	b	c	d	e	f	g	h	g
2	c	d	e	f	g	h	g	h
3	d	e	f	g	h	g	h	g
4	e	f	g	h	g	h	g	h
5	f	g	h	g	h	g	h	g
6	g	h	g	h	g	h	g	h

All the possible initial configurations of a cellular automaton are on the top row and their subsequent evolution after six time steps runs vertically. Note that the number of configurations diminishes so that only g and h are final states. Note that in this context we are abusing the word state to also mean configuration. At the first time step, the probability

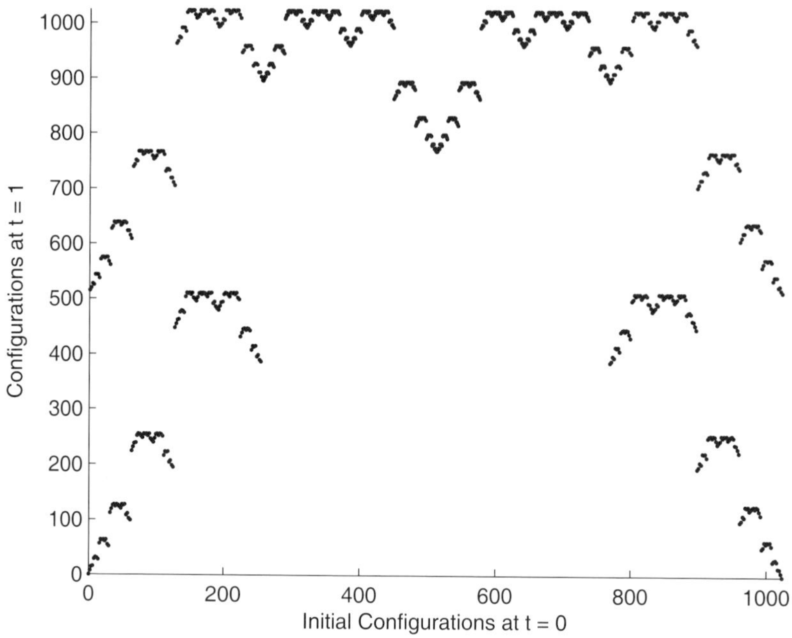

Figure 3.25 A single time step evolution of Rule 126 for each of the 1024 initial configurations of 10 cells wide. Note the the various plateaus of dots indicating the 217 favored states after 1 time step. The graph is also symmetric about the centerline owing to the fact that it has a symmetric rulestring: $126 = 01111110$.

of occurrence of configuration g is $\frac{1}{4}$ and that of h is $\frac{1}{8}$, whereas at the sixth time step, the probability of occurrence of configurations g and h is $\frac{1}{2}$.

In this simplified example, we find that as the system evolves a is an unreachable configuration. A configuration that is unreachable from the evolution of any initial configuration of a particular cellular automaton is called a *Garden of Eden* configuration and this topic will be pursued more rigorously in Section 4.1.6.

Taking a real example that nicely illustrates this compression of states as the automaton evolves is the graph of the evolution of Rule 126. Considering all the $2^{10} = 1024$ initial configurations that are 10 cells wide, we find that the number of favored states decreases to 217 after just a single time step (Fig. 3.25). By examining the time evolutionary history of all the initial configurations, we find that 124 evolve to fixed

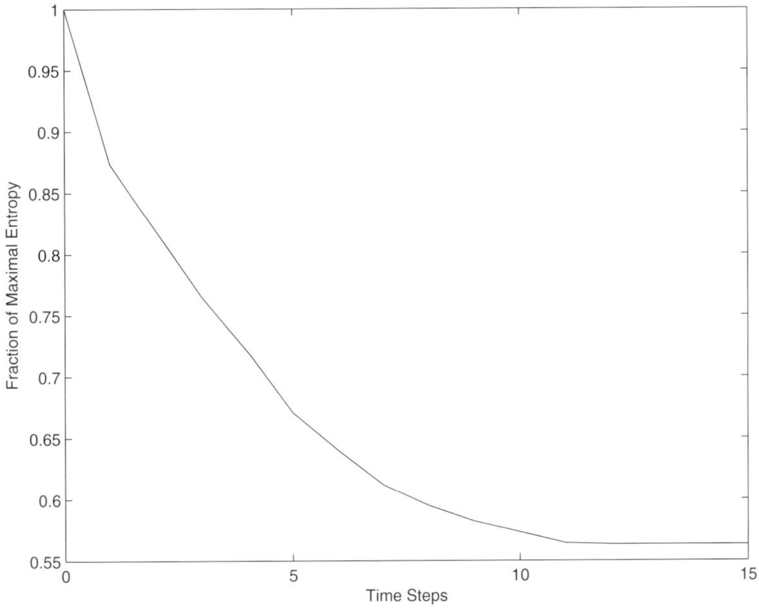

Figure 3.26 Graph showing the entropy as a function of time for the ensemble of 2^{15} initial configurations (cell width = 15) evolving according to (irreversible) Rule 94. The entropy decreases from a maximum at its initial state to a minimum after 14 time steps indicative of irreversibility.

state, 100 become period 2 attractors (alternating between two states), 520 become period 4 attractors, and the remaining 280 become period 6 attractors. Many of these attractors happen to be the same, so that only 81 of them are unique.

In order to compute the entropy of a finite automaton as above, we use the formulation $S = \sum_{i=1}^{2N} P_i \log_2 (1/P_i)$, where P_i is the probability of occurrence of the ith configuration at a given time step as described above. If we compute S at each of various time steps, we typically obtain the result illustrated in Fig. 3.26, showing a decrease over time from maximum entropy that is indicative of the increasing self-organization of the automaton.

If we apply the same procedure in computing the entropy for Rule 30, which in Section 3.5 indicated complete randomness along its central column, we find something rather interesting. Starting from a random

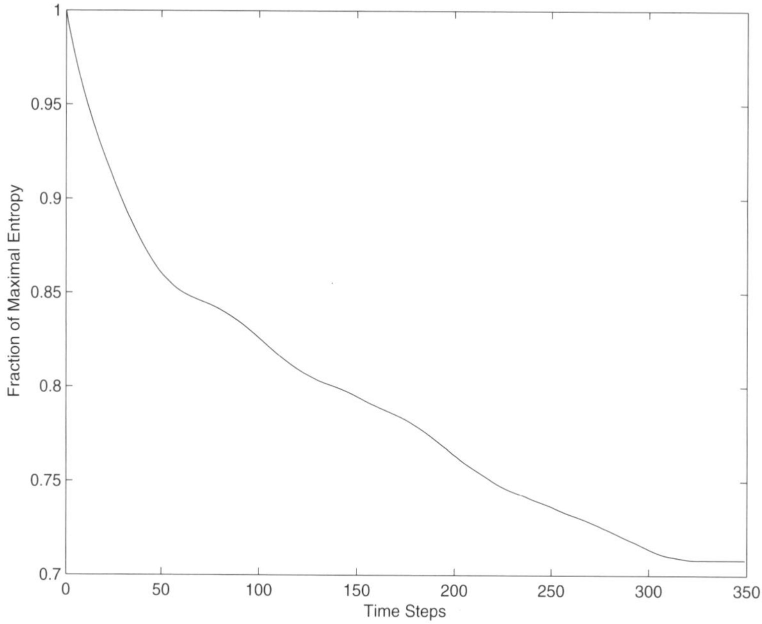

Figure 3.27 Computing the entropy of Rule 30 as it evolves with time we find that it decreases somewhat, indicating a certain degree of overall self-organization (evident in the Rule 30 configuration) even though any individual column is random.

initial configuration, the entropy over time still decreases somewhat, although to a lesser extent than for Rule 126, indicating some degree of overall self-organization (Fig. 3.27). This is in spite of the fact that the black and white pattern in any individual column is completely random!

On the other hand, reversibility in cellular automata is very uncommon but is prevalent in a few elementary cellular automata, namely Rules 15, 51, 85, 170, 204, and 240. As in the case with Rule 15 with random initial conditions (Fig. 3.28) whose transition rule is given by

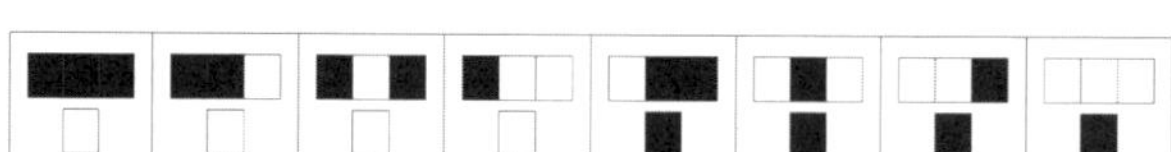

,

Figure 3.28 The reversible CA Rule 15 evolved using random initial conditions.

one can start on any line and go backward in time steps using the rule rotated by 180 degrees, that is,

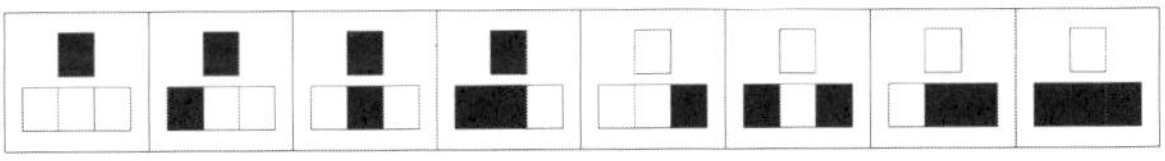

,

which turns out to be Rule 85 (and hence also reversible). Next, in order to reverse Rule 204,

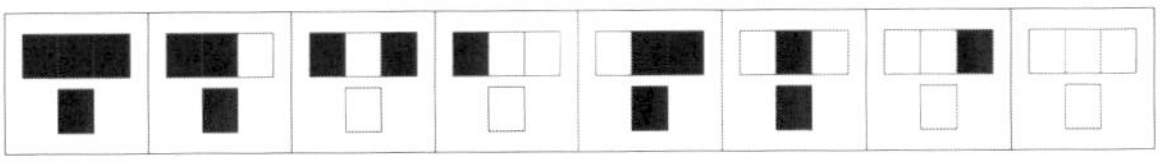

,

its inverse rule

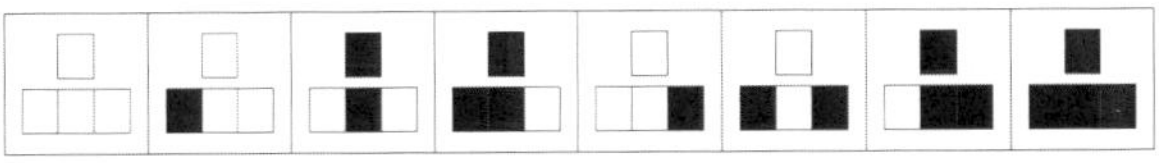

,

is actually itself and the same is true for Rule 51. That just leaves Rule 170 whose inverse is Rule 240.

A mathematical framework for the notion of reversibility is considered later in the sequel concerning the Garden of Eden.

We found that in irreversible automata, the number of configurations could diminish with time (compression of states); however, it is a theorem of the great French mathematician, Joseph Liouville, that the total number of configurations in a reversible system remains constant with time.

Indeed, let us take the reversible CA, Rule 15, and consider the time evolution of the $2^{10} = 1024$ configurations that are 10 cells wide as we did with the irreversible Rule 126. In this instance, it turns out that there are two configurations that become fixed-point attractors, two become period 2 attractors, 30 are period 5 attractors, and the remaining 990 configurations have period 10. As all the attractors happen to be factors of 10, every initial configuration gives rise to itself after 10 time steps, thereby making the whole CA period 10.

On the other hand, Rule 126 differs from Rule 15 in that Rule 126 takes some time before settling into one of its attractor cycles, and indeed, the initial configuration, in general, *never reappears again* for the time evolution of that individual state. In contrast, in the time evolution of Rule 15 all initial configurations reappear at time steps 10, 20, 30,..., for all initial configurations, and thus it is interesting to observe the interplay of the phenomena of incompressibility of states and periodic cycles in a game of "musical configurations."

There is a clever way in which to create *reversible* rules due to Edward Fredkin. What one does is to consider the transition function for the particular rule, say it is given by $\varphi\left[c_{i-1}(t), c_i(t), c_{i+1}(t)\right]$, and then add the state of the central cell at the previous time step: $c_i\left(t-1\right)$. This gives the new rule

$$c_i(t+1) = \varphi\left[c_{i-1}(t), c_i(t), c_{i+1}(t)\right] + c_i\left(t-1\right) \mod 2,$$

which has a unique inverse given by

$$c_i(t-1) = \varphi\left[c_{i-1}(t), c_i(t), c_{i+1}(t)\right] + c_i\left(t+1\right) \mod 2.$$

For example, Rule 90 can be described by

$$c_i(t+1) = c_{i-1}(t) + c_{i+1}(t) \mod 2,$$

which normally is not reversible, but it does have a reversible counterpart, Rule 90R, which can be expressed as

$$c_i(t+1) = c_{i-1}(t) + c_{i+1}(t) + c_i\left(t-1\right) \mod 2.$$

In other words, to determine the value of a cell at the next time step, look to the immediate left of the cell, look to the right, look behind the cell, and take the sum of these three cell states and divide by 2. The remainder, either 0 or 1, is the cell's state at the next time step.

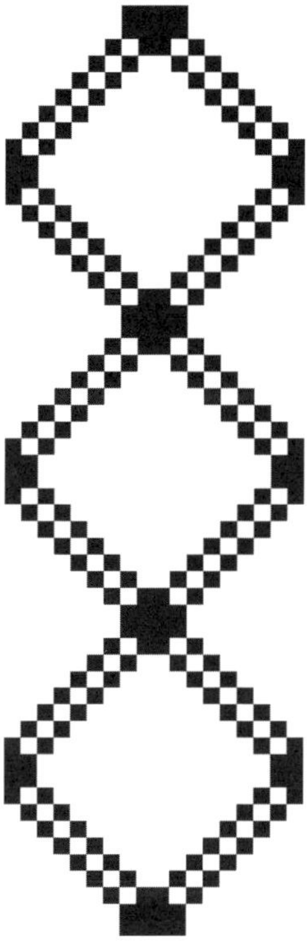

Figure 3.29 This is the reversible Rule 90R with periodic boundary conditions.

The inverse then will have the form

$$c_i(t-1) = c_{i-1}(t) + c_{i+1}(t) + c_i\,(t+1) \mod 2.$$

The evolution of Rule 90R depicted in Fig. 3.29 with periodic boundary conditions can be visually checked and seen to be fully reversible according to the above rules.

The above reversible rules are examples of *second order* cellular automata because a cell's state is a function of the value of the neighborhood's cell states not only at the previous time step, but also at the time step before that. The reversible rule in Fig. 3.24 is Rule 94R.

Cellular automata with more states per cell also have some that are reversible, but as in the case of the elementary ones, it is only a small percentage of the total that actually are.

Figure 3.30 This is Rule 249 depicting the evolution to a uniformly constant state from random initial conditions. Any other set of initial conditions would evolve to the same fixed state.

3.8 CLASSIFICATION OF CELLULAR AUTOMATA

In the early 1980s, Wolfram began to classify cellular automata qualitatively by the complexity of their time evolution behavior. Four basic classes were distinguished, although some automata have a mixture of different classes and some fall between two classes. Nevertheless, the vast majority of automata can be classified in this fashion, from elementary automata to more complex systems. In some respects, this classification parallels the types of behavior exhibited by dynamical systems discussed in Chapter 1 and found in random Boolean networks discussed in Chapter 6. The same four fundamental classes of behavior are to be found over a broad range of cell state values (k), neighborhood radius (r), and dimension of the cellular automaton.

Class I: This behavior is the simplest and evolves to a uniformly constant state as in Rule 249 (Fig. 3.30).

This class of automaton resembles dynamical systems that tend to a fixed-point attractor. In fact, the configuration of all black cells in

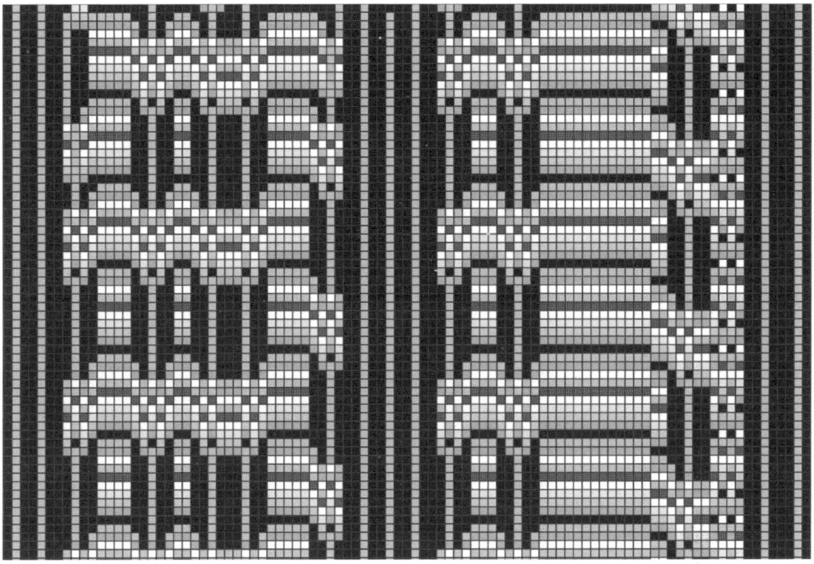

Figure 3.31 This automaton exemplifies the typical behavior of periodic structures found in Class II systems. Found by Mirek Wojtowicz.

Fig. 3.30 is a fixed-point (stable, stationary) attractor in the space of all possible configurations, in that for all initial configurations of the cells, the automaton evolves to this one fixed state. Automata in this class cannot be reversible as the initial information is completely lost.

Class II: In this case, the evolution is toward continually repeating or periodic structures (Fig. 3.31). Class II automata display the periodic behavior exhibited by dynamical systems. They can also act as a kind of filter in that certain sequences of data can be preserved while others are made void. For example, the cellular automaton in Fig. 3.32 preferentially selects the sequence 010 (white-black-white) while all other sequences of 1's and 0's are lost, becoming 0. This sort of selection process is useful in digital image processing.

Related to Class II systems are systems of *finite size*. In these systems, the domain of cells is finite in extent, which is what we have in all practical cases, and this inevitably leads to periodic behavior. In the example below of Fig. 3.33, Rule 150 shows differing periodicities depending on the size of the domain. Periodic boundary conditions are to be used. Because of the finiteness of the domain, the system eventually finds itself in a previous state and thus is forced to repeat itself.

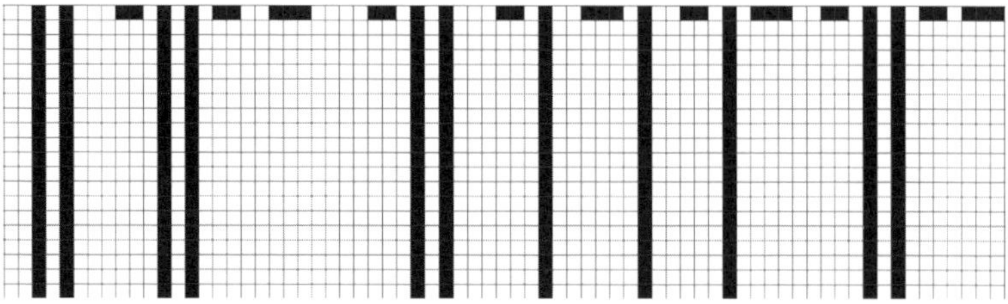

Figure 3.32 This is an example of the preservation of the data string 010 by elementary Rule 4, a Class II automaton.

Figure 3.33 Rule 150 in finite domains (with periodic boundaries) of widths 17, 19, and 31 with periods 15, 510, and 31, respectively.

The maximum possible period is of course affected by the maximum possible number of states, which in the case of two states per cell is 2^n for a domain of n cells. Thus the period can in principle be extremely large even for a modest-sized domain. Often, however, the period is much less than the maximum and its actual value is greatly affected by the domain size.

In a Class II system, as in the preceding example, various parts evolve independently of one another without any long-range communication, thus producing the periodic behavior typical of a system of finite size.

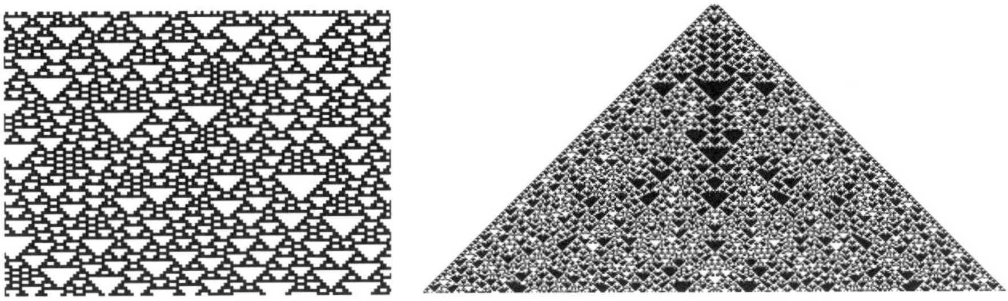

Figure 3.34 This is Rule 126 (left) and Code 2013 showing the typical random behavior found in Class III cellular automata.

Class III: These automata exhibit random behavior typically with triangular features present (Fig. 3.34).

Class III automata are analogous to chaotic dynamical systems. Like the latter, they are very sensitive to initial conditions and play an important role in the study of randomness. For example, Fig. 3.35 shows Class III, Rule 126, with the value of just one cell changed in the initial conditions.

Class IV: In this, the most interesting of all the Wolfram classes but not as rigorously defined as the others, localized structures are produced that move about and interact with each other as the automaton evolves (see Fig. 3.36).

Although Classes I, II, and III demonstrate the basic types of behavior found in dynamical systems, Class IV behavior has no dynamical system equivalent and in fact fits between Class II and Class III in the sense that it is poised between periodicity and randomness. According to Wolfram [1984 and 1994], this is where universal computation could be possible — in the netherworld between order and chaos. This region is the so-called edge of chaos, a term coined by Norman Packard, although the idea probably originates with Christopher Langton [1986]. The salient feature of Class IV behavior is that information can be transmitted through time and space opening up the possibility for complex computations to be made.

Of the elementary one-dimensional cellular automata, we have already mentioned that Rule 110 is capable of universal computation and is of Class IV. By trivially reversing left and right we obtain Rule 124, which of course is also of Class IV, as are the black and white reversals

Figure 3.35 The evolution of Rule 126 from initial conditions differing by only one cell's state. This is typical of Class III behavior and chaotic dynamical systems.

in Rules 137 and 193. But these all effectively have the same dynamics. Among all the other elementary cellular automata, it is possible that Rule 54 is also of Class IV, but according to Wolfram this is not entirely conclusive.

3.8.1 Langton's Parameter

Because there is at least a formal similarity between the types of behavior exhibited by dynamical systems and cellular automata, Christopher Langton sought a "tunable" parameter as we have seen in the analysis of the one-dimensional dynamical system that could be used in some sense to classify automata. However, it must of necessity only be an approximate classification scheme for most questions about the long term evolution of cellular automata are undecidable. For example, the pattern of (Code 357, $r = 1$, $k = 3$) Fig. 3.37 ceases after 88 steps so that this is a Class I automaton.

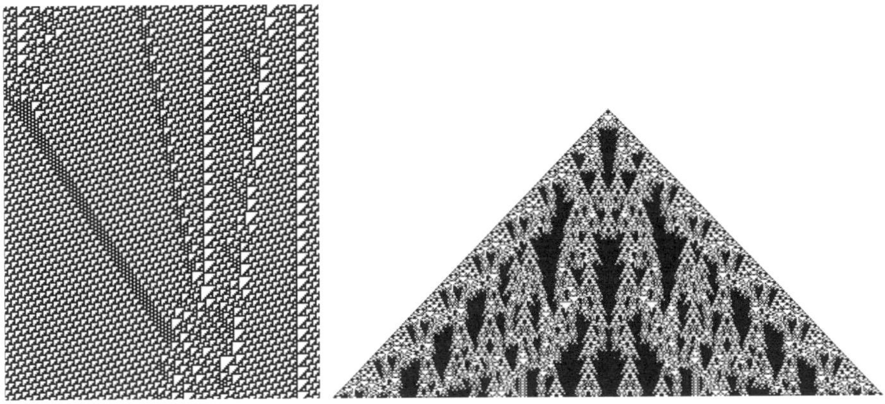

Figure 3.36 Complex interactions between localized structures in Rule 110 and Code 1635 typical of Class IV behavior.

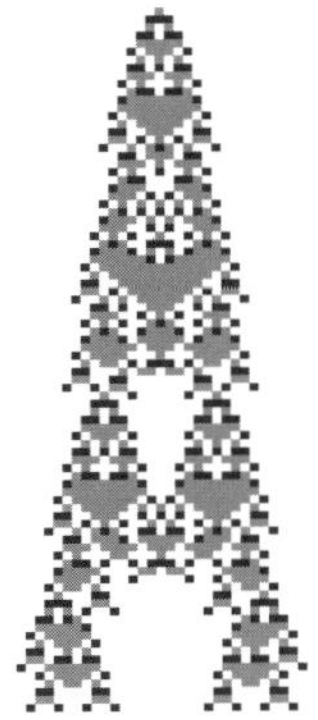

Figure 3.37 The pattern of Code 357, $r = 1$, $k = 3$, no longer evolves after 88 time steps and is thus a Class I automaton.

However, the pattern of another automaton that starts out looking similar to this one carries on indefinitely (Fig. 3.38).

If the behavior dies out quickly, then the automaton is clearly seen to be of Class I, but if it does not, it could still be of Class I, simply reaching a constant state after a very long number of time steps. Or it could be of Class II, with a repeating pattern of possibly a very high period, or even some other class. There is simply no way to determine *a priori* whether the automaton dies out or not other than letting the system evolve for as long as practicable. This question is essentially the same as the halting problem for Turing machines.

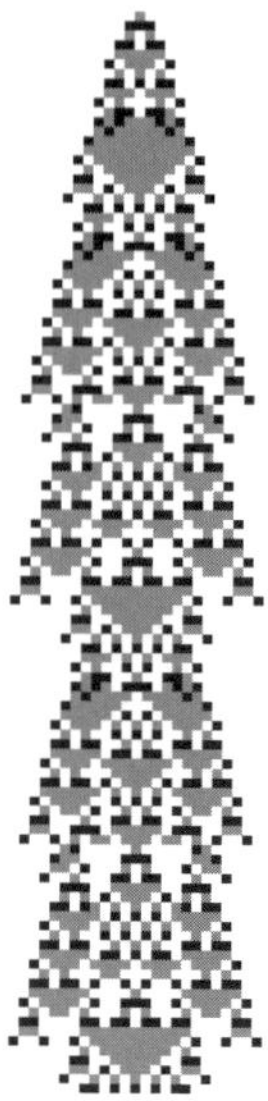

Figure 3.38 Although similar in appearance to the preceding Code 357, Code 600 ($r = 1$, $k = 3$) continues indefinitely as it has a repeating pattern and is thus of Class II.

Fig. 3.39 is another example where first appearances are deceiving. There appears to be the kind of complex interactions that one would expect from a Class IV automaton. However, letting the automaton evolve for some considerable period of time, we find that it settles into typical Class II behavior (Fig. 3.40).

Related to this are cellular automata that Wolfram refers to as "computationally irreducible." There are some physical systems for which it is possible to find a formula or algorithm that will give the state of the system at some time t so that it is not actually necessary to let the system evolve until time t in order to see this. Certainly Rule 255 with its all-black output is an obvious case in point (Fig. 3.41).

If there is to be a formula or algorithm defining the state of a system at a given time t, then it will have to involve less computation than the explicit time evolution of the system itself and hence must be "more sophisticated." But as we will see below, there are some cellular automata (even one-dimensional) that are capable of universal computation and thus able to perform the most sophisticated computations that can be done. The upshot is that there will be some automata for which a description of their long term behavior is not possible and the only way

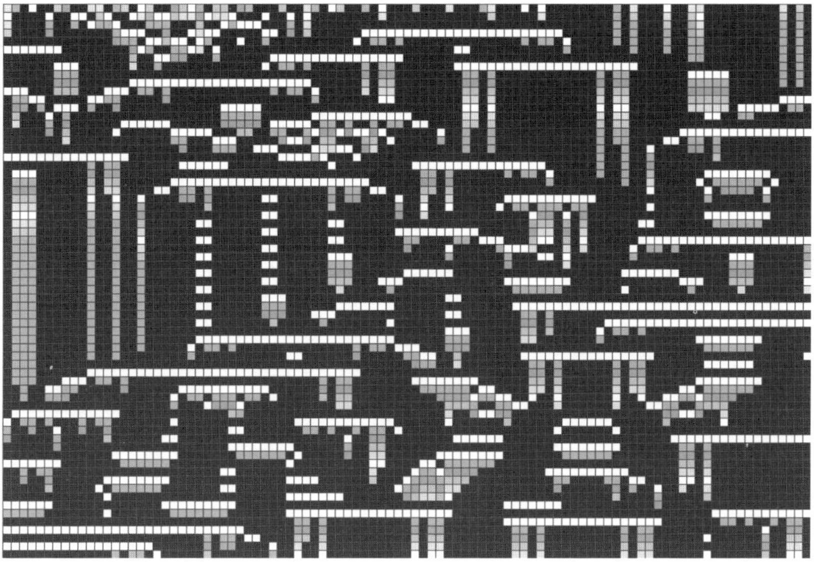

Figure 3.39 This CA apparently looks like Class IV at this stage of its time evolution.

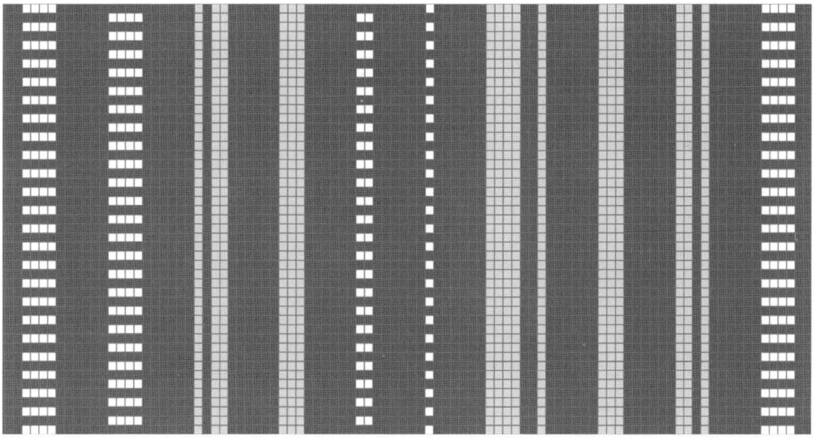

Figure 3.40 The evolution of the cellular automaton in the preceding figure which has evolved to Class II.

to find out is to let them evolve. This can even be the case for Class III automata that are not capable of universal computation such as Rule 30 but whose output is nevertheless random (in the sense we discussed).

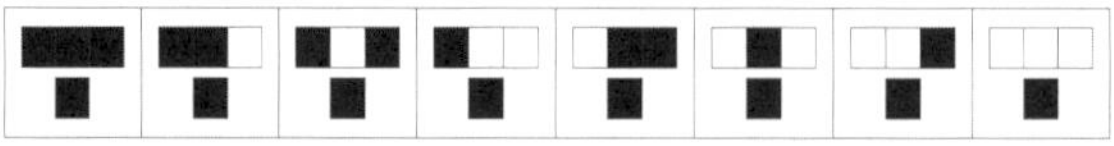

Figure 3.41 Rule 255.

Thus questions about such automata as to whether or not they evolve to a quiescent state are formally undecidable.

In spite of the foregoing difficulties with classification of cellular automata there is still some quantitative measure that we can associate with the four different Wolfram classes. Note that for one particular cellular automaton with k states per cell and a neighborhood radius of r, we found (Section 3.2) there are $n = k^{2r+1}$ different neighborhood-states. For instance, when $k = 2$ and $r = 1$, we have $n = 8$ neighborhood-states making up the transition function. In general, the number of neighborhood-states is given by $n = k^{\rho}$, where ρ is the number of neighbors. If we denote by n_q the number of quiescent states of the central cell (at the next time step) resulting from these neighborhood-states, then the Langton parameter is given by

$$\lambda = \frac{k^{\rho} - n_q}{k^{\rho}},$$

which simply represents the fraction of output states that are active (nonquiescent) in the particular transition function defining the automaton. When $n_q = k^{\rho}$, then all the output states are quiescent and $\lambda = 0$. When $n_q = 0$, there are no quiescent output states and $\lambda = 1$. When all the states are equally represented in the output, we have $\lambda = 1 - 1/k$. So, for example, when $k = 3$ and $r = 1$, there are $3^3 = 27$ neighborhood-states. Denoting the three states by $0, 1$, and 2, and when each occurs with equal frequency in the output, that is, nine times each, two-thirds of the outputs (18 out of 27) will be nonzero (nonquiescent), which equates with $\lambda = 1 - \frac{1}{3}$. In general, for states $0, 1, 2, ..., k - 1$, when each occurs with equal frequency in the output, the proportion of nonquiescent output states is $(k - 1)/k = 1 - 1/k$. Statistically, this is when the output is most heterogeneous and thus the domain of interest is $0 \leq \lambda \leq 1 - 1/k$.

As an example, Rule 5 has $\lambda = \frac{8-6}{8} = 0.25$, which is just the fraction of 1's in the output, and indeed it exhibits the simple periodic behavior of Class II (Fig. 3.42).

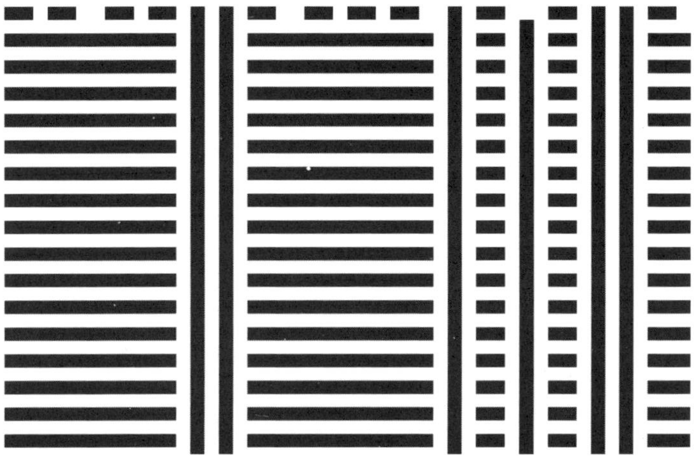

Figure 3.42 Rule 5 displaying Class II behavior with $\lambda = 0.25$.

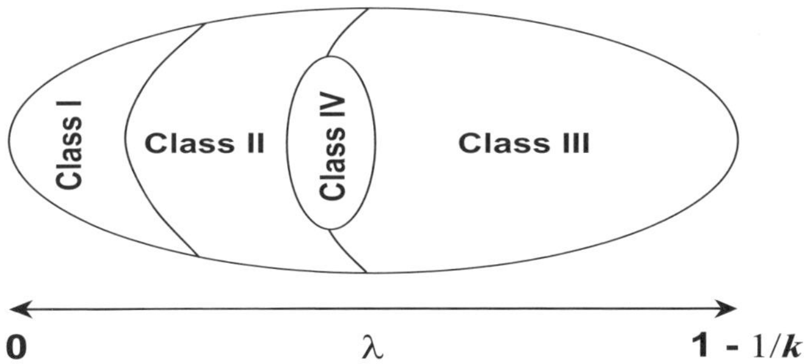

Figure 3.43 The association of the Langton parameter λ with the different Wolfram classes. Note that a specific value of λ can be associated with more than one class.

In Langton's words, the λ value of a transition function "is an aggregrate statistic that is *correlated* with, but not a certain predictor of a certain level of behavioral complexity." It reflects average behavior of a class of rules. Bearing this in mind, we find a correlation with the four classes of cellular automata as indicated in Fig. 3.43.

Here again we observe that Class IV is associated with Class II behavior and precedes Class III. In Class IV, there are periodic structures as in Class II, but unlike Class II, where these structures do not interact, in Class IV automata some of these structures interact in complex ways. Because of these complex interactions, Class IV behavior is also

Figure 3.44 The evolution of Class III, Rule 18 (left), and Class I, Rule 40 (right). Both rules have Langton parameter $\lambda = 0.25$.

on the verge of the random behavior exhibited in Class III automata. This suggests (on average) the following transition of class behavorial regimes with increasing λ from 0 to $1 - 1/k$

$$\text{Fixed Point} \Longrightarrow \text{Periodic} \Longrightarrow \text{Complex} \Longrightarrow \text{Chaotic}$$

This dynamical discrimination works reasonably well for "large" values of k and ρ, say $k \geq 4$, $\rho \geq 5$, but less so for the simplest one-dimensional case of $k = 2$. For example, taking $\rho = 3$, the dynamics are "only roughly correlated" with the Langton parameter. Rule 18,

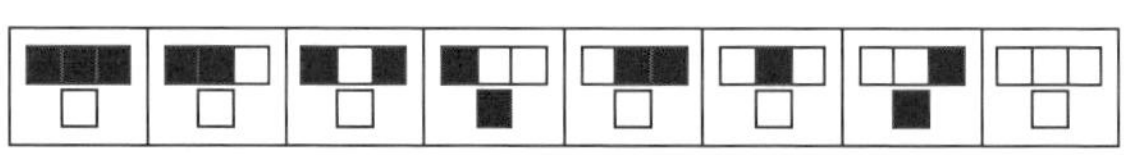

as well as Rule 40,

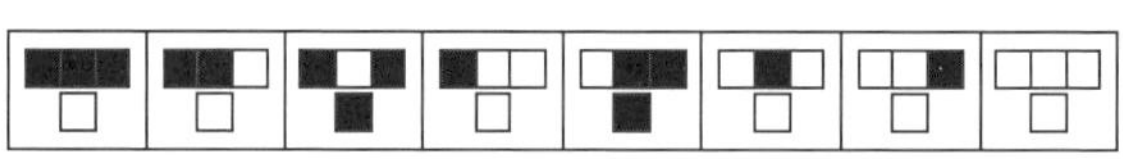

both have $\lambda = 0.25$, yet Rule 18 is Class III (Fig. 3.44, left) and Rule 40 (Fig. 3.44, right) is Class I and we have already seen that Rule 5 ($\lambda = 0.25$) exhibits Class II behavior. Another approach to distinguish the four Wolfram classes, based on what are called *mean field theory curves*, has been taken by H. V. McIntosh [1990].

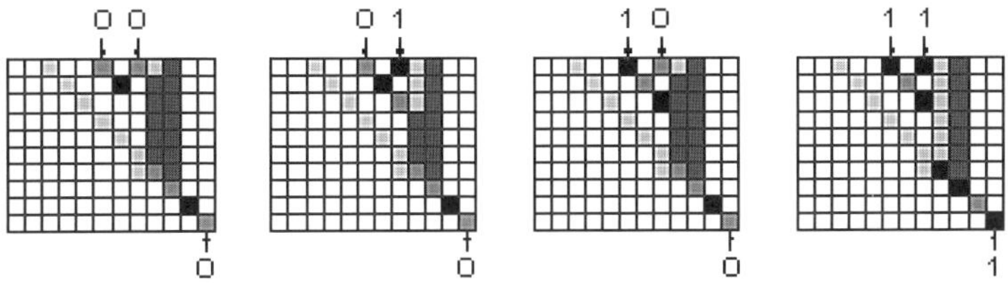

Figure 3.45 The logic AND gate where the input is fed into the first row of the automaton with specific initial conditions and the desired output of 1 at the

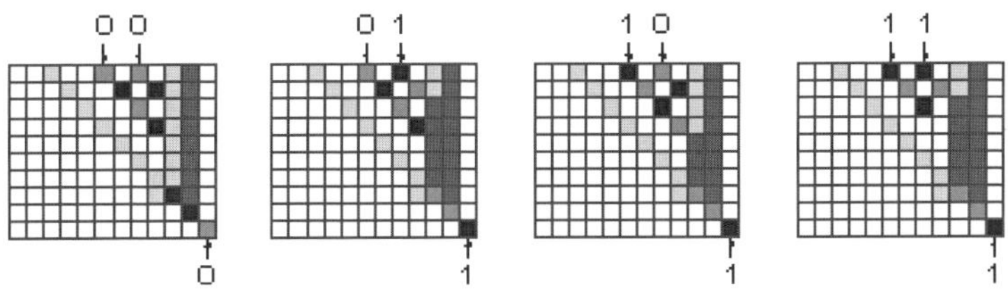

Figure 3.46 The logic OR gate which is a slight variation of the AND gate above giving the output 1 in all but the first case when inputs P and Q are both 0.

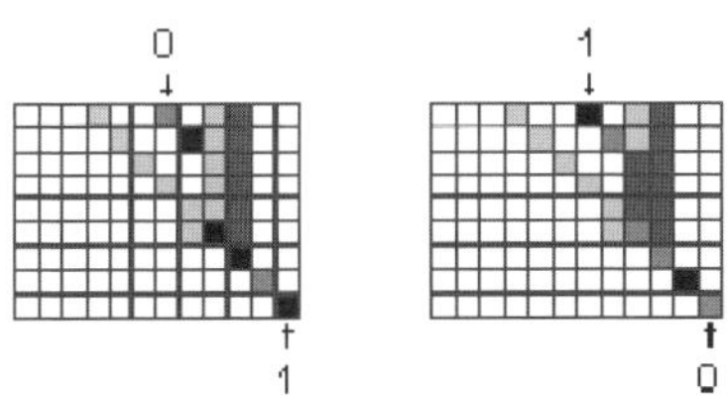

Figure 3.47 The NOT gate turning an input of 0 into 1 and 1 into 0.

3.9 UNIVERSAL COMPUTATION

It is a remarkable fact that one-dimensional cellular automata are capable of universal computation. As mentioned in Chapter 1, this is essentially a consequence of the fact that the logic gates NOT, AND, and OR can be emulated coupled with memory storage and retrieval. These computer functions were demonstrated by Wolfram [2002, p. 662] with the composition of each logic gate given in Figs. 3.45, 3.46, and 3.46. There are five cell states (denoted by five colors — white, light gray, medium gray, dark gray = 0, black = 1) and the transition rule, without going into the precise details, is the same for each of the logic gates.

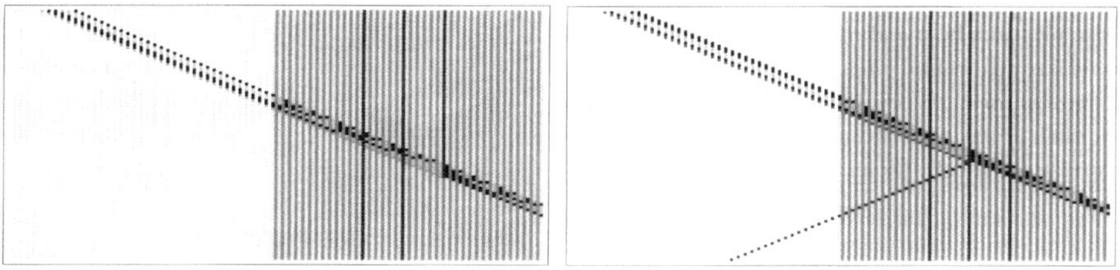

Figure 3.48 A one-dimensional cellular automaton model to illustrate the storage and retrieval of information. The memory is on the right in each figure; in the first figure a 1 is stored in the memory position 13. In the second figure a 1 is retrieved from the memory position 19.

In addition to the logic gates themselves, Wolfram demonstrated how information could be stored and retrieved in random-access memory (Fig. 3.48).

It was Chris Langton's contention [1990], based on his own empirical experiments, that rules capable of complex computations and indeed universal computation are likely to be found for λ values near the transition between order and chaos. This is analogous to Wolfram's contention that Class IV automata are capable of universal computation. The idea being that complex computations require information transmission and storage coupled with interaction between them. The likely place to find such capabilities, according to Langton, is not in the ordered or chaotic regimes but rather in the "phase transition" between them where there are long transients and complex, unpredictable behavior.

Although Langton's intuition apprears to be quite plausible, it is not the whole story. It has been suggested that by suitably "filtering" seemingly Class III automata, it would be possible to uncover within this class CA capable of universal computation (see Mitchell, Hraber, and Crutchfield [1993]). Moreover, one must be rather careful about defining what is meant by "complex computations" which is not necessarily the same thing as "universal computation," the latter perhaps being more what Langton had in mind. But even the ability to achieve universal computation under very specific circumstances within a particular CA environment, while interesting in principle, may be of little use in studying the complex systems found in Nature.

Let us now address a practical problem involving cellular automata and complex computation and see where the chips fall.

3.10 DENSITY PROBLEM

A prototype of many problems in the CA field is that of the so-called density classification problem. In the one-dimensional two-state version it is desired that the evolved cellular automaton becomes either all 1's or all 0's depending on whether the initial configuration was more than half 1's or more than half 0's, respectively. The evolution of such a CA algorithm is thus a means to "classify" the binary string of 0's and 1's according to their frequency (density). In Nature, this is analogous to the phenomenon of *quorum sensing*, whereby certain bacteria are able to coordinate their activities with others of their kind once a certain threshold (quorum) has been reached (see Eshel Ben-Jacob *et al.* [2004]). The problem is actually difficult for a cellular automaton because the global property of discerning the higher frequency of either the 0's or 1's must be carried out by local interactions only. In fact, it was shown by Land and Belew [1995] that the density problem cannot be solved perfectly by any two-state CA having radius $r \geq 1$.

However, all is not lost. If the output conditions of the problem are relaxed slightly, then in fact there is a perfect solution proved by Capcarrere, Sipper, and Tomassini [1996]. The humble elementary Rule 184 given by

$$c_i(t+1) = \begin{cases} c_{i-1}(t) \text{ if } c_i(t) = 0, \\ c_{i+1}(t) \text{ if } c_i(t) = 1, \end{cases}$$

(see also Appendix A for Rule 184's rather unremarkable appearance starting from one black cell) is able to classify any binary string of length N in the following sense. If the density of 1's (respectively 0's) is >0.5, then the automaton evolves to a fixed state with a block (or blocks) of at least two cells in state 1 (respectively state 0) within $[N/2]$ time steps with the rest of the output being alternating 0's and 1's. (Here the brackets mean to round down to the nearest integer.) Even in the case when the density of 1's and 0's is exactly 0.5, the automaton evolves toward a state of solely alternating 1's and 0's, thus perfectly classifiying the density of the initial binary string. The same holds true for Rule 226, which is just the reflected version of Rule 184 (see Fig. 3.49). Here we have a simple example of a phase transition in

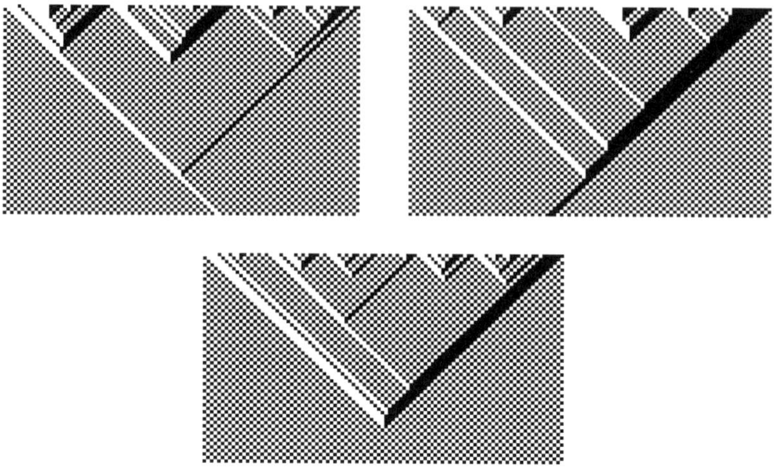

Figure 3.49 Rule 184 classifying densities in the sense mentioned in the text of 51 white/49 black (left), 51 black/49 white (right), and 50 white/50 black (bottom).

a one-dimensional CA, with the variable parameter being the original density.

Since the density problem is not perfectly solvable in its initial formulation, attempts have been made to classify the density of a binary string "most" of the time. One of the best algorithms is the Gaks-Kurdyumov-Levin (GKL) rule, which is nearly always successful with binary strings having high or low initial densities. The GKL rule is a wonderful piece of ingenuity that is a two-state, $r = 3$ rule defined by

$$\begin{aligned} c_i(t+1) &= \text{majority}\{c_i(t), c_{i-1}(t), c_{i-3}(t)\} \quad \text{if } c_i(t) = 0, \\ c_i(t+1) &= \text{majority}\{c_i(t), c_{i+1}(t), c_{i+3}(t)\} \quad \text{if } c_i(t) = 1, \end{aligned}$$

where the term majority means that if two out of the three cells are in state 0 (respectively 1), that value is taken for the central cell update at the next time step.

Roughly speaking, densities of local regions are classified and these regions expand with time. Where there are regions of equal density of 0's and 1's, a checkerboard or white/black boundary signal is propagated indicating that the classification be carried out on a wider scale

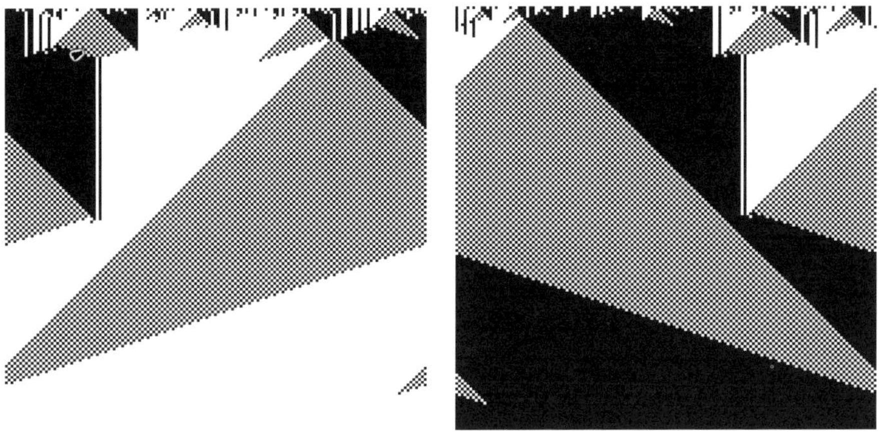

Figure 3.50 The GKL rule classifying initial densities of 77 white/72 black cells (left) and 72 black/77 white (right).

(see Fig. 3.50). As effective as the GKL rule is with initial conditions away from the critical density of 0.5, it was found that for densities very near the critical value the GKL rule failed up to 30% of the time (Mitchell, Hraber, and Crutchfield [1993]).

Norman Packard [1988] experimented with genetic algorithms (see Section 5.4.1) to solve the density classification problem and came to the seemingly reasonable conclusion that when CA are evolved [as in genetic algorithms (GAs)] to perform complex computations, the evolution is driven to rules near the transition to chaos (where in this case the critical Langton parameter values are: $\lambda_c \approx 0.24$ and $\lambda_c \approx 0.83$). Similar experiments conducted by Mitchell *et al.* [1993], also applied genetic algorithms to the density problem to produce very successful evolved CAs, although as it happens, none quite as successful as GKL. The authors' results, however, were "strikingly different" from those reported by Packard. They found that the most successful evolved rules for solving the density problem were found close to $\lambda = 0.5$, which is the theoretical value they also deduced, and *not* near the transition to chaos as Packard had found. Packard's results were seen as possible "artifacts of mechanisms in the particular GA that was used rather than a result of any computational advantage conferred" by the transitional regions.

An evolved CA "capable of sophisticated emergent computation" to solve the density problem is discussed in Mitchell, Crutchfield, and Das [1996], and whose behavior is not unlike that of the GKL rule.

Interestingly, the GKL rule table has $\lambda = 0.5$ (in the middle of the chaotic region), and for what it's worth, Rule 184 also has $\lambda = 0.5$, although it must be borne in mind that the Langton parameter is an aggregrate statistic and an individual CA's behavior can deviate from the average. Nevertheless, as the preceding has shown, it is not necessary that the evolutionary optimal solutions to a complex task have any relation with the Langton parameter normally associated with the so-called edge of chaos.

Again using GAs, an asynchronous approach to the problem was studied by Tomassini and Venzi [2002], although their results showed that the performance of asynchronous algorithms were inferior to the best evolved synchronous CAs or the GKL rule although they were more robust in the presence of noise.

3.11 SYNCHRONIZATION

A challenging problem, of which there are now many variations, first posed by John Myhill in 1957 has to do with the synchronization of the output of an arbitrarily long but finite row of CA. At some time step, all the cells should, for the first time, attain the same state value. Like the density problem, this is difficult for CA since a particular global state must be reached solely by local interactions. This has become known as the firing squad problem or firing squad synchronization problem (FSSP). A cell at one end of the row (a "general") is distinguished from the others (the "soldiers"). Taking a three-cell neighborhood, the problem is to choose the states for each cell and suitable transition functions such that the soldier cells will all be in the same (firing) state for the first time at the same time step. McCarthy and Minsky (see Minsky [1967]) showed that there exist solutions for n cells in $3n-1$, $\frac{5}{2}n-1$,... time steps. A *minimal time solution* is one achieved in $2n-1$ time steps, which is the number of time steps it takes for a general to send a signal to the far-end member and receive a reply. Waksman [1966] gave a minimal time solution involving 16 states per cell, while Balzer [1967] gave one with 8 states per cell and showed that there was no minimal time solution with only 4 states per cell.

The best minimal time solution to date has been demonstrated by Jacques Mazoyer [1987] in a *tour de force* of technical analysis using 6 states (Fig. C.4, see color section). The question is still open whether 6 or 5 states per cell is the least number required. Two generals can also be taken, one at each end, who do not reach the firing state (see examples in Wolfram [2002], p. 1035).

CHAPTER 4

TWO-DIMENSIONAL AUTOMATA

So I have often made the hypothesis that ultimately physics will not require a mathematical statement, that in the end the machinery will be revealed, and the laws will turn out to be simple, like the chequer board with all its apparent complexities.
Richard Feynman – Messenger Lectures, 1964

Two-dimensional cellular automata exhibit some of the same characteristics as do one-dimensional automata. There are two fundamental types of neighborhoods that are mainly considered. First there is the *von Neumann* neighborhood (the five-cell version of which was used in the construction of his self-replicating machine), consisting of the four or five cell array depending on whether or not the central cell is counted (Fig. 4.1).

The *Moore* neighborhood consists of the eight or nine cell array depending on whether or not the central cell is counted (Fig. 4.2).

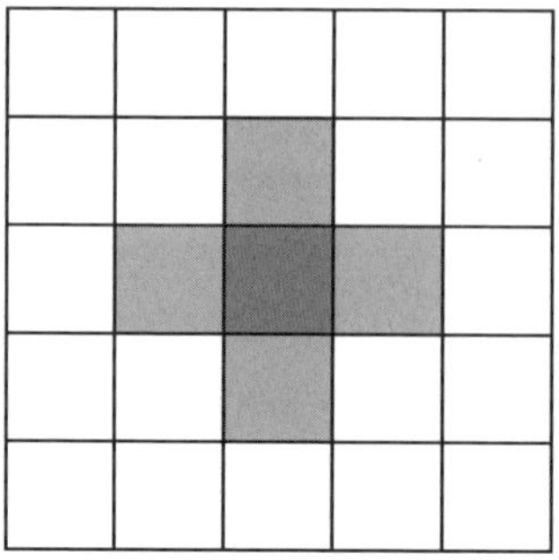

Figure 4.1 The von Neumann neighborhood surrounding a central cell.

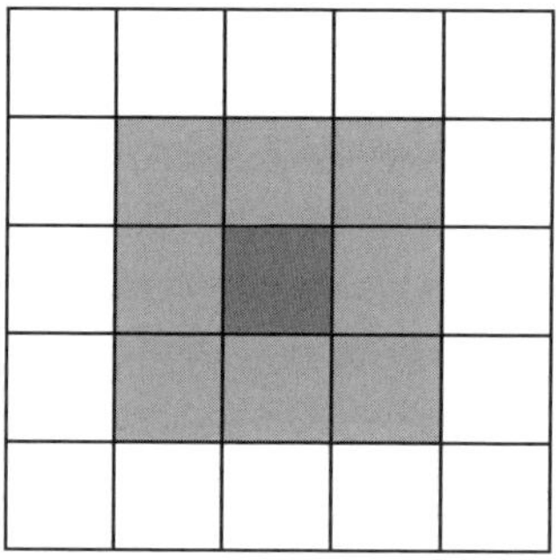

Figure 4.2 The Moore neighborhood with $r = 1$.

In both cases $r = 1$ and each is useful depending on the context. The extended Moore neighborhood has the same form as the preceding but with $r > 1$.

Typically, in a rectangular array, a neighborhood is enumerated as in the von Neumann neighborhood illustrated in Fig. 4.3 and analogously for the Moore neighborhood. The state of the (i, j)th cell at time step t is denoted by $c_{i,j}(t)$.

In the one-dimensional case with $k = 2, r = 1$, there were just $2^3 = 8$ possible neighborhood-states. Now however, with just two states 0 and 1 and a nine-cell Moore neighborhood (again, $k = 2$, $r = 1$), there are $2^9 = 512$ possible neighborhood-states ranging from all white to all

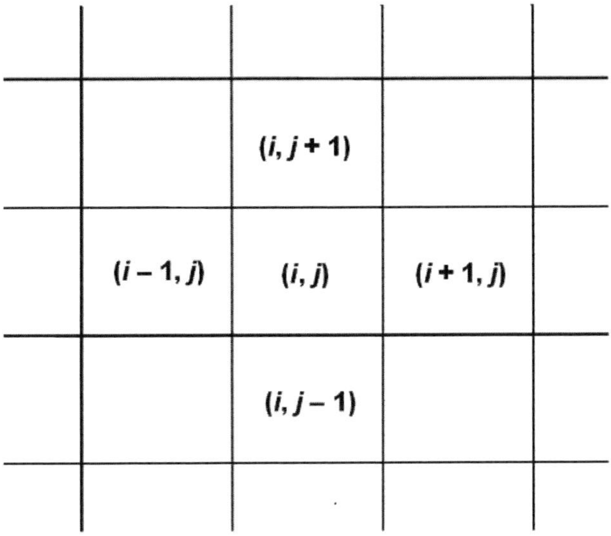

Figure 4.3 The enumeration of the cells of the von Neumann neighborhood.

black with all the various 510 other combinations of white and black cells in between. A given transition function would tell us how the central cell of each of the 512 neighborhood-states should change at the next time step. How many such transition functions are there? A staggering $2^{512} \approx 10^{154}$, more than the number of all the atoms in the universe! Even with a five-cell neighborhood, there are still $2^{32} \approx 10$ billion possible transition functions to choose from.

Beginning in the early 1960s, Stanislaw Ulam and co-workers J. Holladay and Robert Schrandt at Los Alamos Scientific Laboratory began using computing machines to investigate various two-dimensional cellular automata. An infinite plane was considered and divided up into identical squares. The transition rules were eclectic and the results were mostly empirical. One particular automaton was as follows: A cell became alive at the next generation if it was a neighbor (in the von Neumann sense) of exactly one live cell of the current generation. This rule was later coupled with a "death rule" that required all cells that were a fixed number (m) of generations old to die. Say if $m = 2$, then the $(n + 1)$th generation is derived from the nth generation and the $(n - 1)$th generation is erased. Thus only the last two generations survive in any configuration (see Fig. 4.4).

Figure 4.4 The "dog bone" configuration as referred to by Schrandt and Ulam [1967] in which a cell becomes alive in the next generation if it has exactly one alive (black) neighbor in the current generation and any cell of the previous generation dies.

We can use the same transition rule together with the death rule and two different initial configurations on the same array, say in different colors, to distinguish their respective live cells. The growth rule prohibits the appearance of a live cell at the next generation that has two live neighbors at the present generation, and this condition now includes neighbors of either color. This leads to contests between the two opposing systems with the result that one or both may end up eliminated (Fig. 4.5).

Figure 4.5 A contest on a finite array following the rules given in the text, in which the forces of darkness on the top right of the first image gradually eliminate the forces of lightness (gray) on the bottom left. Solid cells represent the current generation, crosses the previous generation. The images represent the situation at generations 11, 32, 49, and 77, respectively.

4.1 THE GAME OF LIFE

> *It is probable, given a large enough "Life" space, initially in a random state, that after a long time, intelligent self-reproducing animals will emerge and populate some parts of the space.*
>
> — *John Conway – Winning Ways for Your Mathematical Plays*

The Game of Life first entered the world stage from the pages of Martin Gardner's *Mathematical Games* column in the October 1970 issue of *Scientific American*. The creator of the Game of Life was the English mathematician John Horton Conway. Originally, Life was to be played out using counters on a chess or Go board, but soon computer screens became Life's natural domain. The Game of Life caused an international sensation following its rather dramatic creation over a period of two years (R. K. Guy in [1985]):

> ... only after the rejection of many patterns, triangular and hexagonal lattices as well as square ones, and of many other laws of birth and death, including the introduction of two and even three sexes. Acres of squared paper were covered, and he and his admiring entourage of graduate students shuffled poker chips, foreign coins, cowrie shells, Go stones or whatever came to hand, until there was a viable balance between life and death.

In the end, the rules for the Game of Life turned out to be quite simple as each cell has exactly two states (1 – alive, or 0 – dead) with the eight-cell Moore neighborhood being the one considered to determine the next state of the central cell:

- A dead cell becomes alive at the next generation if exactly three of its eight neighbors are alive.
- A live cell at the next generation remains alive if either two or three of its eight neighbors are alive but otherwise it dies.

In anthropomorphic terms, the second rule says that if a cell is alive but only one if its eight neighbors is also alive, then the first cell will die of loneliness. On the other hand, if more than three of a cell's neighbors are also alive, then the cell will die of overcrowding. By the first rule, live cells are born from a *ménage á trois*. Of course, this is not really a game that you play in the conventional sense but rather a microcosm of another universe that one can explore since we know its physics entirely. An excellent exposition of the Game of Life can be found in the monograph *The Recursive Universe* by William Poundstone [1985].

There are many variants of Conway's original set of rules, as well as using other lattices such as triangular or hexagonal ones or considering Life in other dimensions, but none seem to offer the richness and diversity of the original game. It is common to let the Game of Life evolve on a lattice with periodic boundary conditions ("wrap"), that is to say, cells on the extreme left and right are considered neighbors and cells at the extreme top and bottom of the lattice are also to be considered neighbors. Whereas using periodic boundary conditions in the one-dimensional case resulted in a continuous loop, in the two-dimensional case our lattice becomes a torus (donut shaped).

The Game of Life is an example of a Class IV automaton. We note that Packard and Wolfram [1985] found no two-dimensional Class IV cellular automata other than "trivial variants of Life." When the updating of cells is asynchronous (see Section 4.4), the Game of Life no longer exhibits Class IV behavior but instead converges to a stationary state (Bersini and Detour [1994]).

Life is also an example of an "outer totalistic" rule in that the state of the central cell at the next time step depends on the prior state of the central cell as well as the *sum* of the values of the eight neighboring cells. The preceding rules for the Game of Life can be represented in the following transition table:

$c_{i,j}(t)$	sum								
	8	7	6	5	4	3	2	1	0
0	0	0	0	0	0	1	0	0	0
1	0	0	0	0	0	1	1	0	0

where the row that one considers for the value of $c_{i,j}(t+1)$ is determined by either the 0 or 1 value of the central cell state $c_{i,j}(t)$. So, for example, if the central cell $c_{i,j}(t)$ has the value 1 and the sum of the states of the eight neighboring cells is 3, then $c_{i,j}(t+1) = 1$.

Conway had considered the rules very carefully so that Life is delicately poised between having patterns that die out quickly and those that continue to grow. In fact, when Life was first proposed, Conway conjectured that no initial collection of live cells could grow without bound and offered a $50 prize to the first person who could prove or disprove this before the end of 1970. The conjecture was short lived as we will see below.

4.1.1 Lifeforms

There is an entire pantheon of "lifeforms" and the interested reader can refer to Eric Weisstein's Treasure Trove website at: `http://www.ericweisstein.\\com/encyclopedias/life/`. We will discuss the evolution of just some of the more common ones. Clearly any fewer than three cells will die in one generation. For a triplet of live cells (that do not vanish after the first generation), they either rapidly evolve to extinction (top three rows of Fig. 4.6), become a block of static cells that remains unchanged with all subsequent generations (fourth

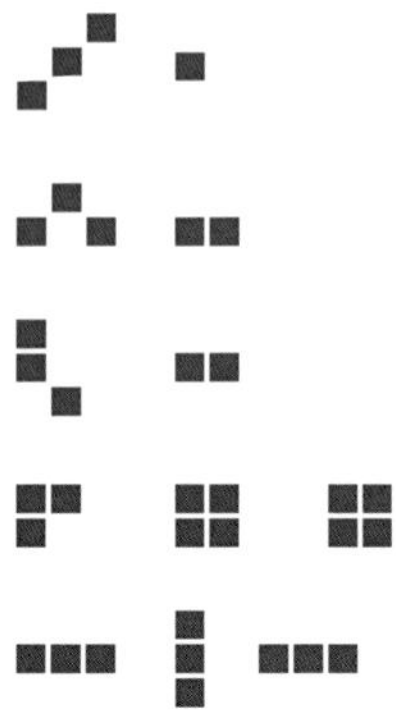

Figure 4.6 The possible evolutionary histories of three cells in the Game of Life. Here time evolution is horizontal. The orientation of the initial cells is not relevant.

Figure 4.7 The evolution of 4 live cells with time increasing to the right. The last two configurations of the last row alternate in a two-cycle.

row), or become an oscillating two-cycle triplet (bottom row). Four-cell configurations evolve to stable forms (top four rows of Fig. 4.7) as well as a long sequence of various forms.

The five-cell "R-pentomino" (see Fig. 4.19, right) is one of the most fascinating elementary lifeforms in that its evolutionary history does not stabilize until 1103 generations. In the process it generates what are known as "gliders," (discussed below) among other lifeforms.

Figure 4.8 Some common invariant forms (left to right): block, tub, beehive, ship, snake, pond, fishhook or eater, loaf.

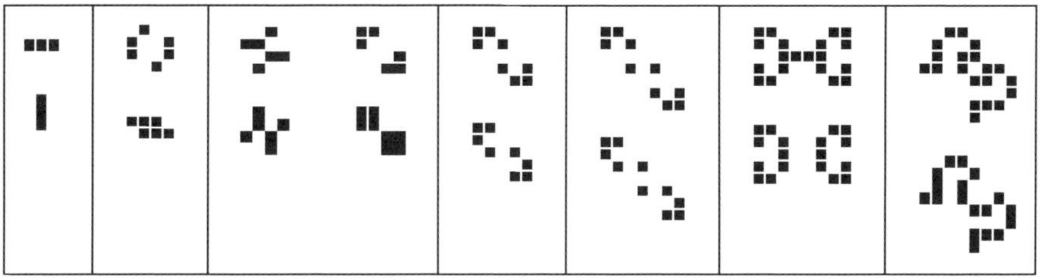

Figure 4.9 Period 2 oscillators. The two rows indicate the two different forms of each oscillator.

4.1.2 Invariant Forms

Some configurations in the Game of Life remain unchanged at every time step (Fig. 4.8). We have already seen the block of four cells and what is known as the "beehive" (the terminal figure of rows 2, 3, and 4 of Fig. 4.7). The "block" is the most common and it turns up frequently in Life. As each cell of the block has exactly three live neighbors, they all persist in time yet surrounding dead cells have only two live neighbors, which is insufficient for them to spring into life.

4.1.3 Oscillators

There are some lifeforms that exhibit periodic behavior oscillating indefinitely between a fixed number of configurations. Period 2 oscillators alternate between two distinct states; they arise spontaneously and are very common; see Fig. 4.9.

A variety of other cyclic oscillators (e.g., Figs. 4.10 and 4.11) have been created artificially, including periods 3, 4, 5, ..., 18, including one of period 144 by Achim Flammenkamp.

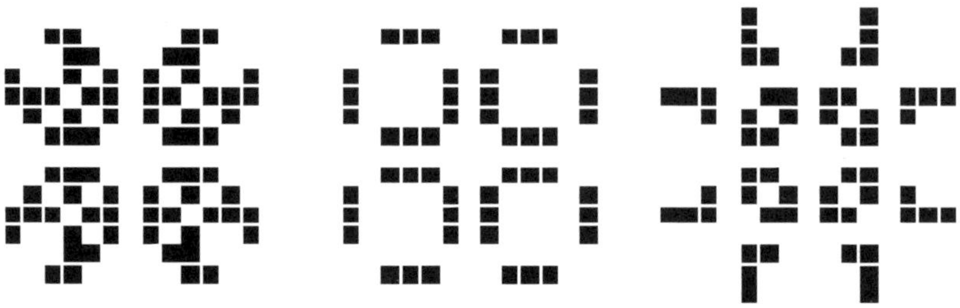

Figure 4.10 An elaborate period 3 oscillator known as the CP-pulsar displaying its three different states. The final state returns to the first at the next time step.

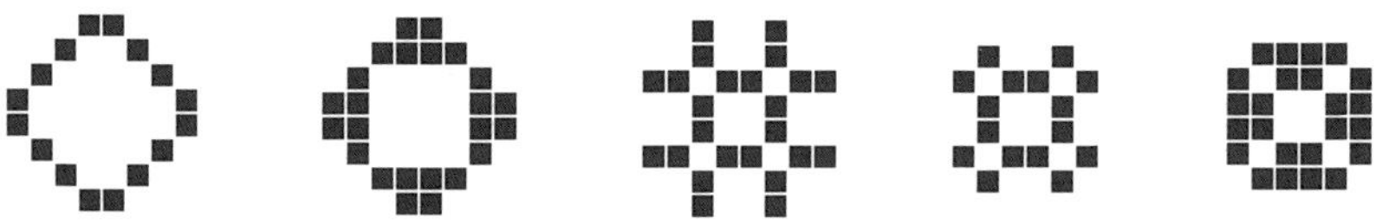

Figure 4.11 A period 5 oscillator displaying its five different lifeforms.

4.1.4 Gliders

Gliders are another five-cell configuration and they actually move one cell diagonally at the fourth time step (Fig. 4.12). They are known as gliders as at time step $t + 2$, they are reflected in a diagonal line, mathematically, a "glide reflection." By time step $t + 4$ the glider is reflected once again back to its original orientation, but one cell (diagonally) displaced. This 4-cycle is then endlessly repeated. The glider is a marvelous creature to watch as it marches in its ungainly fashion across the computer screen. In the words of computer scientist Steve Grand [2001], "The glider is a thing — a coherent persistent phenomenon that moves across 'space'—and yet is not separate from or superimposed on that space. It is simply a self-propagating disturbance in the space created by these little rule-following [cells]" (p. 40).

The maximum speed that information can propagate from one cell to another (horizontally, vertically, or diagonally) is one cell per generation. In Life, this is known as the *speed of light* (c). Since the glider moves exactly one diagonal cell after four generations, it is said to move at one-fourth the speed of light ($c/4$).

Figure 4.12 A glider moves one cell diagonally to the right after four generations.

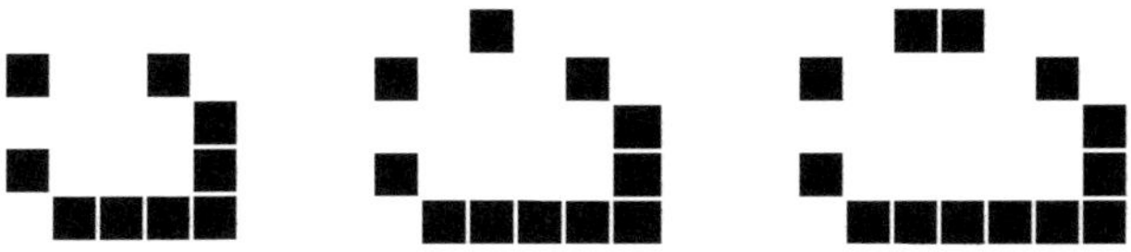

Figure 4.13 From left to right: lightweight, medium weight, and heavy weight spaceships. These move horizontally at the speed $c/2$.

Conway has proved that the maximum speed of any configuration moving horizontally or vertically is $c/2$. Configurations that actually do move at this speed are what Conway called "spaceships," depicted in Fig. 4.13.

One of the most remarkable configurations to arise in the early days of Life was the "glider gun." This arose out of work done by Robert April, Michael Beeler, R. William Gosper, Jr., Richard Howell, Richard Schroeppel, and Michael Speciner, who were in the Artificial Intelligence Project at M.I.T. In November 1970 they claimed the \$50 prize offered by Conway by demonstrating the glider gun (Figs. 4.14 and 4.15) found by Gosper that would indefinitely generate gliders every 30 generations, thus disproving Conway's conjecture that the number of live cells cannot grow without bound. Somewhat, remarkably, Gosper's group found that the collision of 13 specially arranged gliders can create their glider gun. Since then glider guns have been found with other periods of generation, even one having period 256.

Another way to produce unbounded growth in Life is via a "puffer train." These objects travel in a vertical direction and leave behind "smoke" or "debris" that becomes stable. The first puffer train was also found by Bill Gosper and consisted of an engine escorted by two

Figure 4.14 The initial configuration of the original glider gun discovered by Bill Gosper that generates a new glider every 30 generations.

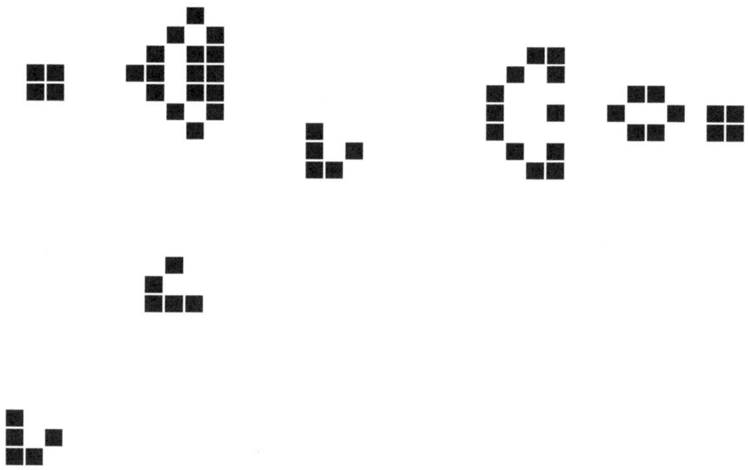

Figure 4.15 The glider gun after it has fired off three gliders toward the lower left.

lightweight spaceships. Since then numerous other ones have been discovered (Fig. 4.16).

There are also "glider eaters" that devour gliders and are useful in the creation of logic gates (Fig. 4.17).

Another type of unbounded growth was discovered by David Bell in 1993. Starting with a finite configuration of live cells, it produces a "spacefiller" which is a feature that can appear in other cellular automata (Fig. 4.1.4).

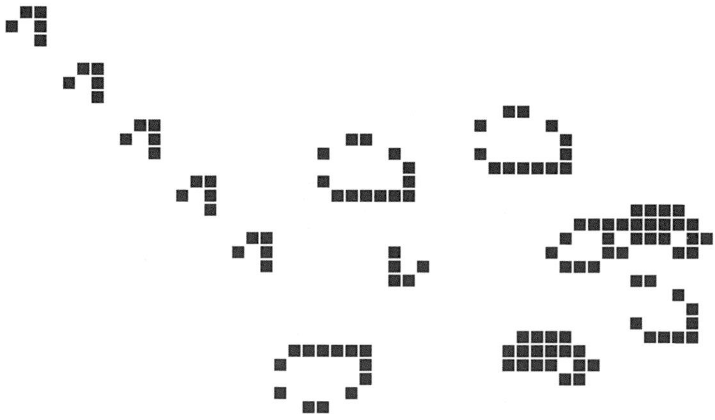

Figure 4.16 A period 16 puffer train (at right) that produces a smoke trail. Based on a design of Tim Coe.

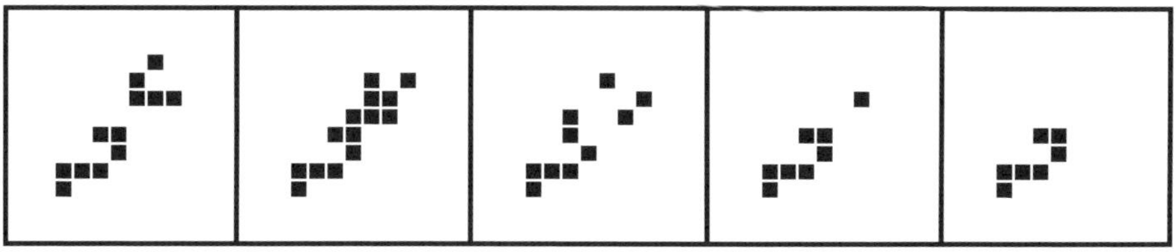

Figure 4.17 In this sequence, a glider eater in the bottom left of the first frame is confronted by a glider approaching at 45 degrees. The glider is gradually eaten until it has disappeared in the last frame. The glider eater is an essential component of certain Game of Life logic gates.

4.1.5 Methuselah Configurations

Such initial patterns have less than 10 live starting cells but they continue to evolve to considerable old age before stabilizing and necessarily exclude configurations that grow forever in the sense of an ever-increasing number of alive cells. An R-pentomino (Fig. 4.19 right) remains alive for 1103 generations having produced six gliders that march off to infinity. The acorn (center) was discovered by Charles Corderman and remains alive for 5206 generations. Rabbits were discovered by Andrew Trevorrow in 1986 and stabilize after 17,331 into an oscillating 2-cycle having produced 39 gliders.

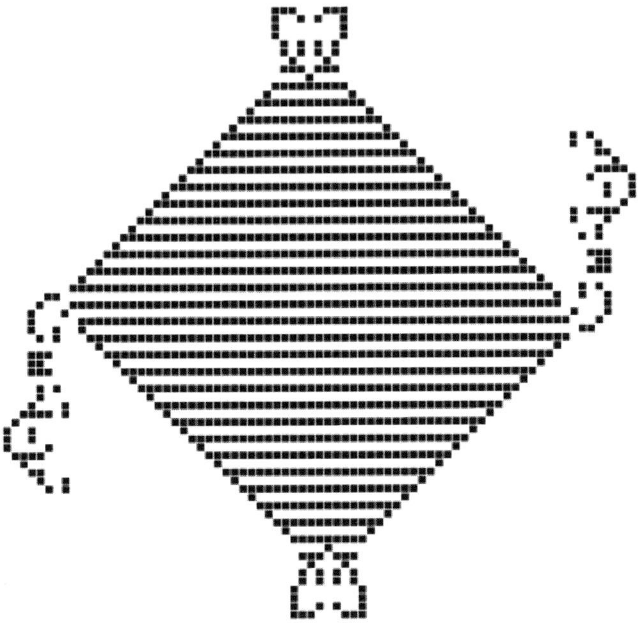

Figure 4.18 The spacefiller lifeform of David Bell after 46 time steps.

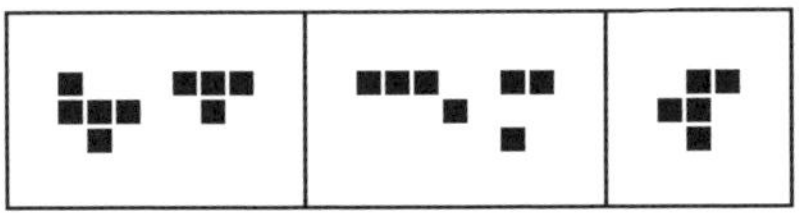

Figure 4.19 Methuselah configurations (left to right): rabbits, acorn, and R-pentomino.

4.1.6 Garden of Eden

The evolution of a cellular automaton is governed by the local transition function which alters the states of all cells in the array synchronously at discrete time steps. Thus, patterns of cells are changed into other patterns of cells as the system evolves. It is natural to ask if there are some patterns of cells that do not arise at all in the evolution of the system, except as initial conditions?

A cellular automaton configuration that can have no prior configuration generating it (via the underlying local transition function) is called a *Garden of Eden* pattern, a term due to John W. Tukey of Princeton University. In a paper in 1962, Edward F. Moore found that in the evolution of a cellular automaton, if a particular configuration had more than one distinct predecessor, then there would have to exist some configuration that would have no predecessor (the Garden of Eden pattern). This was only an "existence theorem" and no method was given for actually finding the Garden of Eden configuration. The converse to this result was established by John Myhill in 1963, namely, if there is a Garden of Eden configuration with no predecessor, then some configuration must have two distinct predecessors. Both results are quite general, applying to any cellular automata of finite configurations (i.e., those configurations with only a finite number of nonquiescent cells — although the configurations are infinite in number) and in any dimension. It is clear that if the array consists of a *finite number* of cells, say N, then the total number of possible configurations allowing for two possible states per cell is just 2^N, with at most 2^N possible outcomes resulting from the transition function acting on each of these. If it so happens that two different configurations are transformed by the action of the transition function onto the same configuration, then only $2^N - 1$ configurations have been generated by the transition function, leaving one configuration without a precedessor. But the Moore and Myhill theorems are about the *infinitely* many finite configurations of cellular automata within infinite arrays.

Because the Game of Life does have configurations that have more than one predecessor (for example, the block and beehive have multiple precedessors as is seen in Figs. 4.6 and 4.7), and the fact that finite configurations are transformed to finite configurations, Garden of Eden patterns must exist and indeed have been found over the years. Roger Banks found the first one in 1971 with 226 live cells. The one with the smallest number of live cells (at the time of this writing) is 143 due to Achim Flammenkamp (Fig. 4.20, right).

The preceding considerations can be discussed in a more mathematical framework. In general, a function f is called *injective* (or *one-to-one*) if whenever $x \neq y$, then $f(x) \neq f(y)$. In other words, distinct

Figure 4.20 The Garden of Eden configurations for the Game of Life of Roger Banks (left) and Achim Flammenkamp (right), having 226 and 143 live cells respectively .

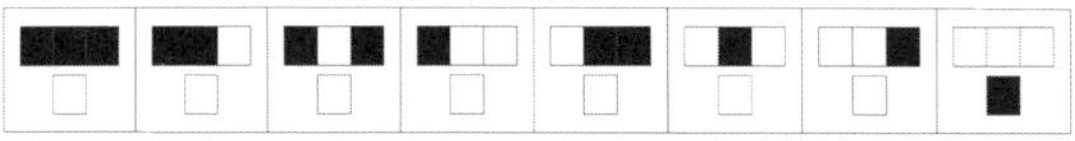

Figure 4.21 Rule 1 allows something to be created from nothing, which is not allowed in the present context.

points x and y must have distinct *images* $f(x)$ and $f(y)$. This also means that if $f(x) = z$, then no other point can be mapped by f to z, since if some other point y did satisfy $f(y) = z$, then we would have $x \neq y$ and $f(x) = f(y)$, a clear violation of the injective property of f.

We now consider the collection of all *finite* configurations C of a cellular automaton, such as the Game of Life. By finite, we mean that all but a finite number of cells in the lattice are in the quiescent state. Then any configuration belonging to C is transformed by the local transition function acting on each cell into another configuration and, moreover, that configuration will also belong to C if we make the supposition that if a cell and all its neighbors are in a quiescent state, then the cell remains quiescent at the next time step. Therefore, although the array is infinite in extent, only finitely many cells at any time step ever leave the nonquiescent state and no infinite configuration ever becomes the successor of any finite configuration. We have already encountered an example in Rule 1 where this is not the case since its rulestring is given in Fig. 4.21.

In this instance a quiescent neighborhood generates a live central cell at the next time step. This results in an initial finite configuration of one black cell generating an infinite configuration at the next time step (see Fig. 3.14), and it is exactly this sort of behavior we wish to exclude. We are in a sense getting something from nothing and that is not allowed here.

The transformation from one configuration into another at each time step induced by the local action of the transition function can be considered as another function, the *global transition function* F that maps configurations c belonging to C to other configurations c' in C. So, for example, in the one-dimensional case taking nearest neighbors, if ϕ is the local transition function, we define the global transition function by

$$[F(c)]_i = \phi[c_{i-1}, c_i, c_{i+1}],$$

where we have suppressed the advance of one time step in the notation.

If the global transition function F is *not* injective, then this would mean that two distinct configurations, say c and c' in C, would be mapped to the same configuration (remembering that F is induced by the local transition function), so that there would be a configuration with two predecessors (c and c') as was discussed above. Therefore, having more than one predecessor configuration is equivalent to the global function F not being injective.

Since F is a mapping from C to C, symbolically, $F : C \longrightarrow C$, we can ask if *every* configuration c in C arises as the image of the global transition function acting on some other c belonging to C? If the answer is yes, then we say that the function F is *surjective* (or *onto*). In other words, for every c' in C, there is some other c in C such that $F(c) = c'$. Then, if it so happens that F is *not* surjective, we find that there is some c' belonging to C with no other configuration in C that is transformed into it. Such a configuration c' is then a Garden of Eden configuration since it would have no precedessor arising from the global transition function F and thus from the local transition function. Therefore, the existence of a Garden of Eden configuration is equivalent to the global transition function being not surjective.

We can now state:

Theorem 1 (Moore-Myhill Theorem) *If C is the collection of all finite configurations and $F : C \longrightarrow C$ is the global transition function, then F is not injective if and only if it is not surjective.*

In other words, if some configuration has more than one precedessor (F is not injective), then there is some configuration with no precedessor

(F is not surjective), and conversely. Moore's theorem is the first result and Myhill's the converse. In most contexts injective and surjective are two mathematical properties that normally have nothing to do with one another.

The global transition function (in the preceding context) being injective is now also surjective and this is equivalent to the cellular automaton being *reversible* (also called *invertible*), that is, for every configuration it is possible to go back to a unique predecessor at each time step, as was discussed in the section on reversibility in Chapter 3. Since we know that the Game of Life has Garden of Eden configurations, then by the Moore-Myhill theorem it cannot be reversible. But, of course, we already know this as, for example, the beehive and block configurations have more than one predecessor. In general, given the local transition function for a two-dimensional cellular automaton, the question of reversibility is undecidable (Kari [1990]), although for one-dimensional cellular automata, the question of reversibility is decidable (Amoroso and Patt [1972]). For an interesting survey of invertible CA, see Toffoli and Margolus [1990].

When Moore [1962] proved his theorem, he was interested in the question of when a machine was unable to replicate itself. About the Garden of Eden configuration, Moore had this to say. "Since it is a machine that cannot arise by reproduction, it must be a machine that cannot reproduce itself." Another result of Moore's about the rate at which a cellular automaton configuration can replicate itself is found later, in Section 4.3.

It is worth noting, that if we drop the restriction of finite configurations, then in the general case we have the following result of Hedlund [1969], who was one of the earliest pioneers in the subject of cellular automata:

Theorem 2 (Hedlund Theorem) *If the global transition function on the set of all configurations is injective, then it is also surjective.*

The situation is more straightforward with respect to additive CA. If φ is the local transition function acting on cell states $c_1, c_2, ..., c_n$ in some neighborhood given by

$$\varphi(c_1, c_2, ..., c_n) = \sum_{i=1}^{n} \lambda_i c_i \mod m,$$

then the global transition F is surjective if the greatest common divisor of all the numbers $\lambda_1, \lambda_2, ..., \lambda_n, m$, is 1 (see Chaudhuri *et al.* [1997]).

4.1.7 Universal Computation in Life

An outline of a proof that the Game of Life was capable of universal computation was presented by John Conway (in Berlekamp, Conway, and Guy [1982]) and independently by William Gosper. The key here is to use a glider gun to emit gliders at regular intervals. This permits the transmission of information from one region to another and simulates the electrical pulses of a regular computer. As the Game of Life is a Class IV cellular automaton (whose Langton parameter $\lambda = 0.273$, which lies in the phase transition region), it is not surprising that it could be capable of universal computation, in view of Wolfram's conjecture that all Class IV automata should have this capability.

As was mentioned in Chapter 1, at the heart of any computer system is the construction of the logic gates NOT, AND, OR. What Conway and Gosper demonstrated was that each of these logic gates could be constructed within the Game of Life, together with a form of infinite memory storage.

In order to construct the NOT gate we consider that P is either 0 or 1 and a data source gun emits a glider whenever P is 1 and nothing whenever P is 0. An emitter glider gun at E is positioned as in Fig. 4.22 and is synchronized to fire gliders simultaneously. Whenever two gliders collide, they will annihilate one another. Hence, P being 0 permits the unfettered passage of a glider from the gun E, thus turning a 0 into a 1, whereas if P is 1, it emits a glider that collides and annihilates the glider from the gun E, resulting in a void that turns the 1 into a 0. The result at receptor R is *not* P.

We can construct the AND logic gate $P \wedge Q$ by adjoining a bit more structure to the NOT gave as indicated in Fig. 4.23. According to the truth table the conjunction $P \wedge Q$ is to be the value 1 only when P is 1 and Q is 1. In all other cases, $P \wedge Q$ is to be 0. Taking P and Q both to have truth value 1 as in the figure results in a glider emanating from the glider gun Q being annihilated by the glider emanating from the emitter E, whereas the glider from P passes unhindered to produce a 1 at the receptor R. However, if P is 1 and a glider emanates from P, and Q is 0 so that nothing emanates from Q, then the glider from P this time is annihilated by the one from E and the receptor R receives no

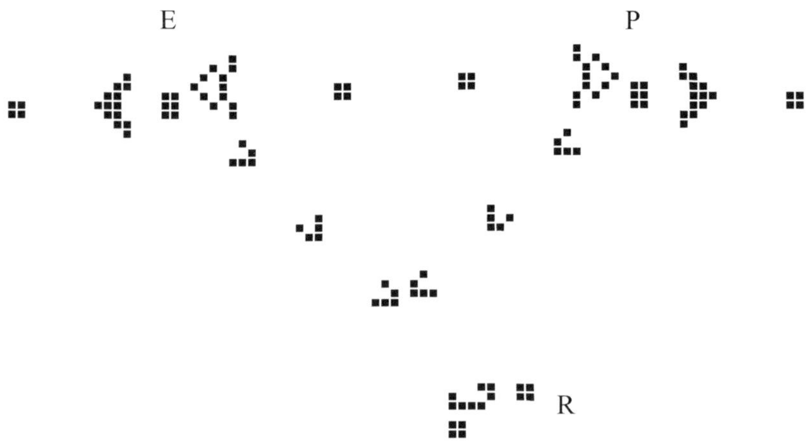

Figure 4.22 The implementation of the NOT gate in the Game of Life. If P is 0, the result at the receptor R is 1, and if P is 1, the receptor receives no input, or 0 since the colliding gliders annihilate one another.

data, hence is 0. On the other hand, if P is 0 and Q is 1, then the glider from Q is annihilated by the glider from E so that again R becomes 0. Finally, if both P and Q are 0, then the glider from the emitter E is eaten by the glider eater and R is once again 0.

The OR gate incorporates a portion of the AND gate as depicted in Fig. 4.24 but now we are provided with two emitter glider guns, both designated E. Note that $P \vee Q$ has truth value 1 if P or Q is 1 (or both) and is 0 in all other cases. Suppose first that P is 0 and Q is 1 as in the figure. Then the gliders from Q are annihilated by the emitter on the right, but the emitter on the left produces a signal at the receptor R. Similarly, if P is 1 and Q is 0, P's gliders are annihilated this time, but the receptor R still receives an input. If both P and Q are on, then again R receives a signal. Only when both P and Q are off do the gliders from each emitter annihilate each other and there is no signal at the receptor.

Memory storage for the universal computer is accomplished by simulating a Minsky register by sliding a Life block utilizing a salvo of gliders to either push it or pull it along a diagonal. In the Conway

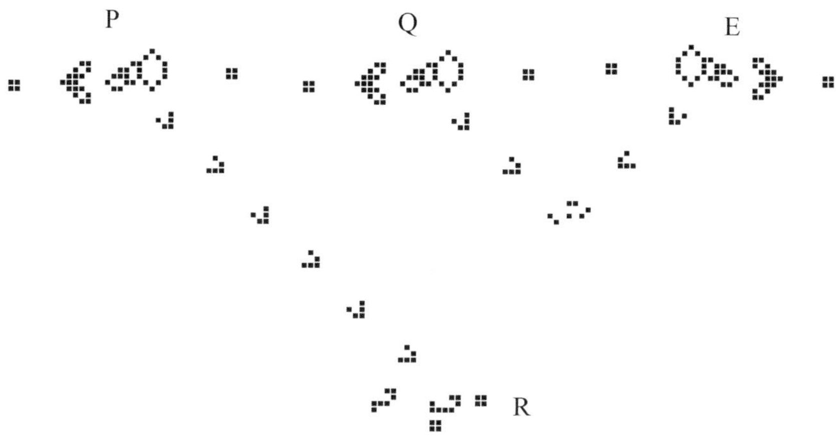

Figure 4.23 The AND gate in the Game of Life. Only when both P and Q are on does the receptor R receive any data. If only P or Q is on (i.e., emitting gliders), then the gliders from E annihilate them and the receptor R receives nothing. Note the glider eater at the bottom left which mops up gliders emitted by E when both P and Q are off.

model, the block (whose location gives its numerical value which can be arbitrarily large) could be pulled three units along a diagonal by a 2-glider salvo, whereas a 30-glider salvo was required to push the block three units along the diagonal. A test for zero was also implemented. This "sliding block memory" was later improved by Dean Hickerson in a much more practical construction that allowed a block to be pulled just one unit by a 2-glider salvo and pushed one unit by a 3-glider salvo. This later became the basis for the construction of a universal register machine implemented in Life by Paul Chapman (`http://www.igblan.free-online.co.uk/igblan/ca/index.html`) in November 2002. This machine is capable of universal computation with communication between the various components carried out by lightweight spaceships. Therefore the Game of Life has all the computational ability of any modern electronic computer.

In this context it should also be mentioned that Paul Rendell in 2000 constructed a Turing machine, however, with a finite tape which could in principle be extended to a universal Turing machine and thus capable of universal computation.

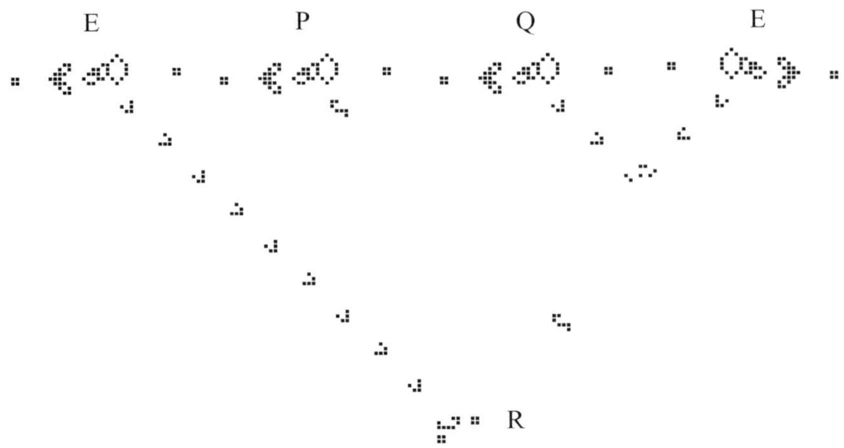

Figure 4.24 The OR gate in the Game of Life. If both P and Q are off, then the gliders from each emitter E annihilate each other and there is no signal at the receptor R. In the other cases when either P or Q are on or both are on, a signal is received by the receptor R from the emitter on the left.

In the spirit of von Neumann, Conway in *Winning Ways* demonstrated that there are configurations in Life capable of self-replication.

There are many websites devoted to the Game of Life. Actually seeing the mélange of lifeforms develop and evolve is a wonderful experience to behold. One of the best implementations is *Life 32* by John Bontes, which can be downloaded at: `http://psoup.math.wisc.\\edu/Life32.html`. An excellent resource website is Eric Weisstein's Treasure Trove website cited previously.

4.2 OTHER AUTOMATA

There is actually a myriad of two-dimensional cellular automata that have been created over the past 30 years. One of the classical models is called *Brian's Brain* and is due to Brian Silverman (Fig. C.5, see color section). The automaton has three cell states denoted as ready (0), firing (1), and refractory (2). The rules bear some resemblance with how neurons in the brain behave (a more sophisticated model of neural activity will be presented in Chapter 5):

- A cell fires only if it is in the ready(0) state and exactly two of its (eight) neighbors are firing(1).
- Upon firing, a cell changes to the refractory state (2) for one time step and then reverts to the ready state (0).

As in the Game of Life, there are various brainforms such as "haulers," ""butterflies," and "twizzlers." The butterflies are the analogue to the gliders in Life and move diagonally at the rate of one cell every four time steps. A "butterfly gun" has been implemented by Rudy Rucker in his CelLab environment.

Another ingenious creation by Silverman is called *Wireworld* and allows for the creation of sophisticated electronic circuits. This automaton has four states: background (0), electron head (1), electron tail (2), and wire (3) and the following set of rules for each time step:

- Background (usually black) cells never change their state.
- Electron head (1) cells change their state to electron tail (2).
- Electron tail (2) cells change their state to wire (3).
- Wire (3) cells change their state to electron head (1) if one or two of its eight neighbors are electron heads (1).

An adjacent pair of an electron tail (T) and electron head (H) comprise an "electron" which can be sent along a strand of "wire" (W) cells in accordance with the preceding rules (THWWW...$\longrightarrow$WTHWW...). With this set-up it is possible to build AND, OR, and NOT gates, memory storage, and hence a computer. An AND gate is depicted in Fig. 4.25.

A simple adder has been constructed by student Peter Lane that computes the sum in base 2 of two inputs as in Fig. 4.26. Here the two inputs are given on the left and fed into the circuit so that the following binary output is computed:

input A	input B	bottom output (2^1)	top ouput (2^0)
0	0	0	0
0	1	0	1
1	0	0	1
1	1	1	0

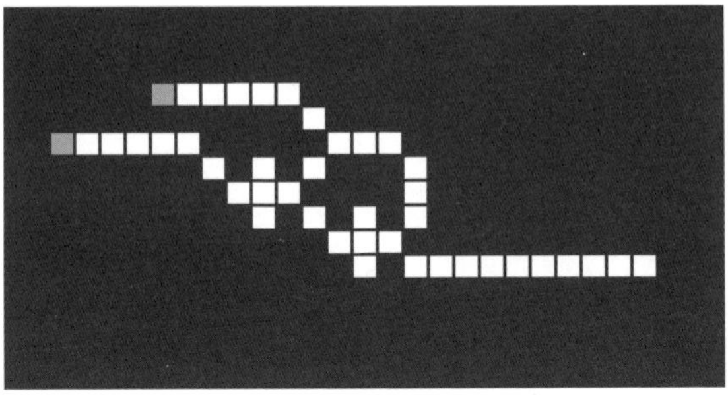

Figure 4.25 The arrangement of the AND gate in *Wireworld.* Two electron head cells are the inputs at the left and move along the paths of wire cells. Only when there are two electrons entering this configuration will the output be an electron. Other cases result in no output.

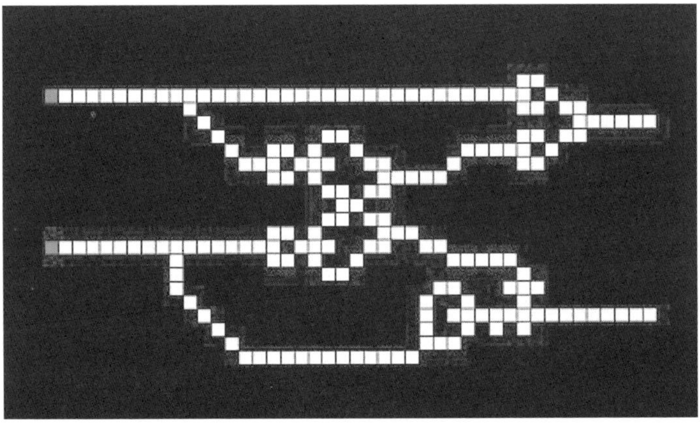

Figure 4.26 A simple adder in *Wireworld* that computes the sum of two inputs at A (top left) and B (bottom left) in base 2 with the result outputed at the top right (2^0) and bottom right (2^1)

A binary adder has also been demonstrated in the Game of Life by D. J. Buckingham [1978].

Another interesting two-state, two-dimensional cellular automaton is called "vote" (also called "majority") created by Gérard Vichniac and is a totalistic rule given by the rule table

sum	9	8	7	6	5	4	3	2	1	0
$c(t+1)$	1	1	1	1	1	0	0	0	0	0

.

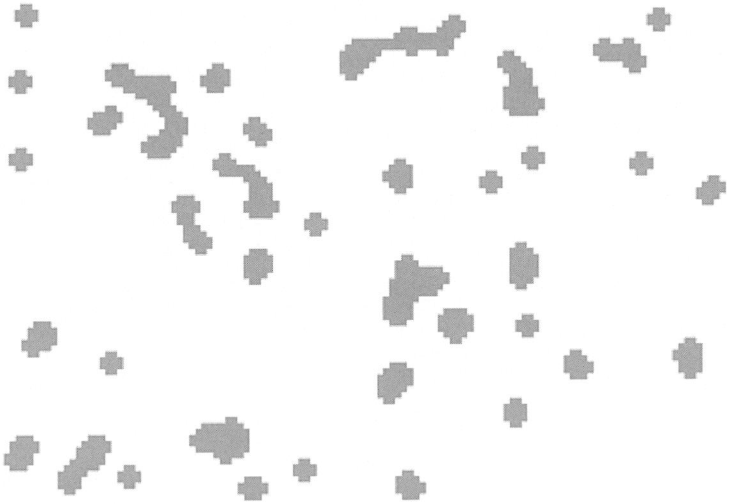

Figure 4.27 The cellular automaton vote which decides the value of the central cell of a Moore neighborhood by what a majority of the values of its neighbors are. The initial configuration had 60% pale gray (zero-state) cells and the above is the equilibrium state.

where $c(t+1)$ is the value taken by the central cell of the nine-cell Moore neighborhood at the next time step. Here one can easily see where the automaton gets its name. If five or more of the cells (i.e., a majority) are 1, then the central cell's state also takes the value 1 at the next time step. But if less than five cells (i.e., a minority) take the value 1, then the central cell becomes 0. The time evolution from a random initial state depends critically on the initial concentration of 1's and 0's and yields large shifting regions made up respectively of the two states and dominance at the initial stage leads to even greater dominance at the equilibrium stage (Fig. 4.27). The totalistic code number is 992, which is 1111100000 in base 2. Von Neumann neighborhoods can also be considered so that $c(t+1)$ becomes 1 whenever the neighborhood sum is 3 or more. These automata are related to models of percolation and pattern formation on animal coats.

There is another more scientific way to depict the vote rule by using the so-called *Heaviside step function H*, also referred to as a *threshold* function (Fig. 4.28). This is defined as follows:

$$H(x) = \begin{cases} 1 \text{ if } x \geq 0, \\ 0 \text{ if } x < 0. \end{cases}$$

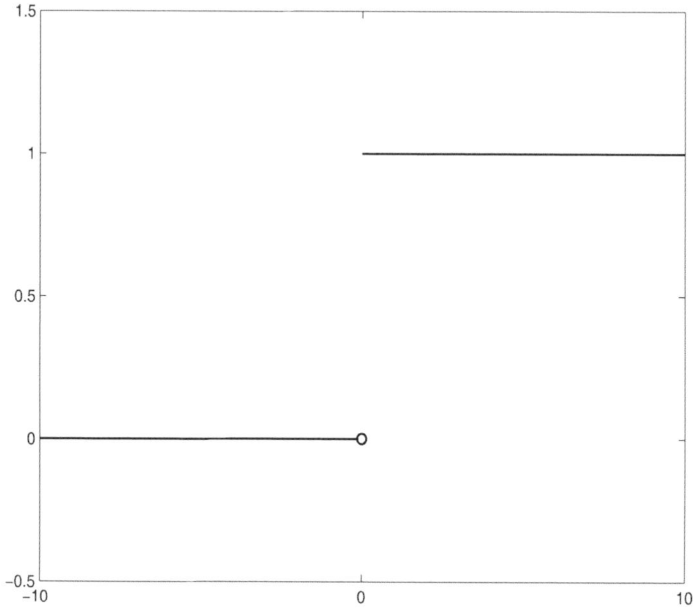

Figure 4.28 The Heaviside step function, which only has a nonzero output when a nonnegative threshold has been reached.

Thus you obtain an output of 1 if the quantity x, whatever it happens to be, is nonnegative, and otherwise the output is 0. If we denote the nine cells of the Moore neighborhood (with a slight abuse of our usual notation) by $c_1(t), c_2(t), \ldots, c_9(t)$, then, according to the vote rule, the value of the central cell at the next time step is given by

$$H\left(\sum_{i=1}^{9} c_i(t) - 5\right),$$

whereby if five or more of the cells have state value 1, then the function H returns the value 1, otherwise it is 0. The value of 5 is just a threshold value that "turns on" the central cell at the next time step once the threshold is reached. This notion is an underlying feature of various cellular automata models in biology (see Section 5.4).

Another variant of the vote automaton due to Vishniac reverses the states of the sum 4 and 5 values.

sum	9	8	7	6	5	4	3	2	1	0
$c(t+1)$	1	1	1	1	0	1	0	0	0	0

This has the effect of permitting interchange at the boundary between regions of opposing colors where the majority is not very strong for either. This CA has served as a model of annealing.

4.2.1 Partitioning Cellular Automata

A new type of neighborhood was devised by Margolus that is fundamentally different from either the von Neumann or Moore neighborhoods. It consists of a partitioning of the usual lattice into blocks of cells which are 2×2 in size in the simplest case and which we only consider here. The transition rule, or rather *block rule*, updates the entire block as a distinct entity rather than any individual cell as in the usual cellular automaton. Another unique feature is that two overlapping partitionings into 2×2 are employed, one given by "dark" lines and one by "light" lines as in Fig. 4.29. At each time step, one switches from one type of neighborhood to the other.

The block rule is applied to all 2×2 blocks alternating between utilizing the dark neighborhoods and light neighborhoods. At first glance this system appears to be very unlike an ordinary cellular automaton, but with more states and neighbors, it can indeed be expressed as an

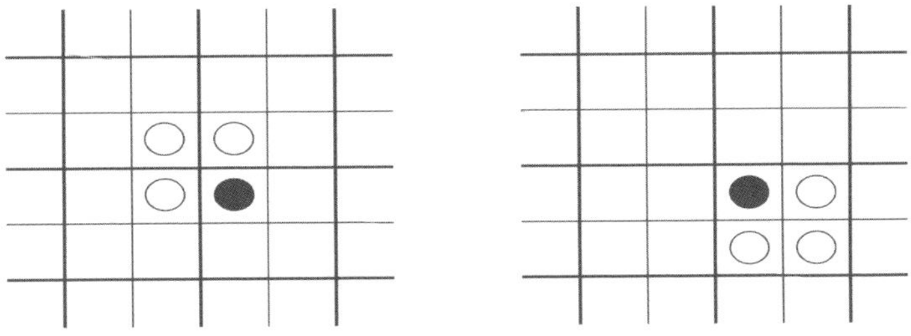

Figure 4.29 A Margolus neighborhood consists of two distinct overlapping 2×2 blocks of cells and a particular cell has successively two overlapping neighborhoods. The neighbors of a black cell (indicated by the black dot) in a light neighborhood are indicated at the left, and the neighbors of a dark neighborhood are as on the right.

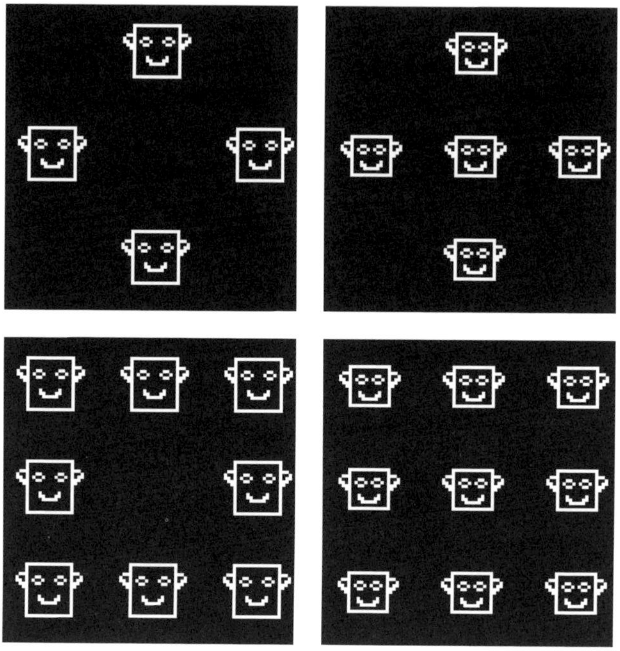

Figure 4.30 The resulting replication after 32 time steps applying the parity rule described in the text to four variations of the type of neighborhood (von Neumann four-cell and five-cell, top row; Moore eight-cell and nine-cell, bottom row).

ordinary cellular automaton. The Margolus neighborhood comes into its own in the section on lattice gas automata in Chapter 5.

4.3 REPLICATION

It is possible to demonstrate a trivial form of self-replication in the two-dimensional cellular automaton model of Edward Fredkin. In our first example, known as the *parity rule*, given a four-cell von Neumann neighborhood:

- A central cell state becomes 1 (alive/black) if it had an odd number of black (0) neighbors at the previous time step.
- A central cell becomes 0 (dead/white) if it had an even number of black (0) neighbors at the previous time step.

This is also the transition function of the one-dimensional Rule 90 and is outer totalistic in that we are only considering the sum of the four neighboring cell states. The effect is a quadrupling of any initial cell pattern (Fig. 4.30, top left) each 2^n generations, with n depending on the original configuration.

Here the outer totalistic rule we have used is given by the transition table

sum	4	3	2	1	0
$c(t+1)$	0	1	0	1	0

where $c(t+1)$ is the value of the central cell and the initial pattern has replicated after $32 = 2^5$ time steps (as do all the others below). However, if we take the sum of the five-cell von Neumann neighborhood, we have the totalistic rule

sum	5	4	3	2	1	0
$c(t+1)$	1	0	1	0	1	0

with a five-fold replication of the initial pattern (Fig. 4.30, top right).

We can also consider the analogous replication with respect to an eight-cell Moore neighborhood and here we implement the Fredkin (outer totalistic) rule given by the transition table

sum	8	7	6	5	4	3	2	1	0
$c(t+1)$	0	1	0	1	0	1	0	1	0

which results in an eight-fold replication (Fig. 4.30, bottom left).

And lastly, we have a nine-cell Moore neighborhood counterpart with the totalistic rule

sum	9	8	7	6	5	4	3	2	1	0
$c(t+1)$	1	0	1	0	1	0	1	0	1	0

which replicates a two-dimensional configuration nine-fold (Fig. 4.30, bottom right). This particular parity rule can be designated Code 682, which is the value in base 2 of the binary digits.

It must be noted, however, that none of the foregoing examples represents the same sort of replication in the von Neumann sense we saw in Chapter 1 as it is purely a consequence of the transition function and not accomplished by the automaton itself.

Edward F. Moore, whom we encountered previously regarding Garden of Eden configurations, also made an interesting observation concerning the rate at which a configuration can replicate. Here, again, as in the discussion of Moore's Garden of Eden theorem, we are only considering finite configurations. Again, we also require the further condition that if a cell is in a quiescent state and all its neighbors upon which the transition function acts are in the same state, then the cell remains in the quiescent state at the next time step. If we denote a finite configuration by c and the number of copies of c at any time step t by $\#c$, Moore showed that

$$\#c \leq kt^2,$$

where k is some positive constant. One consequence of this is that a self-reproducing configuration cannot keep doubling with each time step, since the values of t^2 are $1, 4, 9, 16, 25, 36, 49...$ and doubling yields $\#c$ values of $2, 4, 8, 16, 32, 64, 128, \ldots$.

To see how Moore established this result, suppose that the smallest *square* array that contains the initial configuration c has s cells on a side, and hence an area of s^2. At the next time step, $t = 1$, the square array can only grow (as a consequence of the action of the transition function) by one cell on each of its four sides, so that the square array now has area $(s + 2)^2$. At time step $t = 2$, the square array will have an area equal to $(s + 4)^2$, and in general, the array will have area $(s + 2t)^2$ at each time step t. This value also represents the maximum number of alive (non-quiescent) cells at each time step t. Letting a denote the number of alive cells of the original configuration c, the number of copies of c at each time step is at most

$$\frac{(s + 2t)^2}{a} \leq \left(\frac{s^2 + 4s + 4}{a}\right) t^2,$$

and we arrive at Moore's inequality above.

4.4 ASYNCHRONOUS UPDATING

In general, it has been found that the asynchronous cellular automata evolve much differently from their synchronous counterparts. For example, cyclic dynamics can only occur with synchronous updating, although the set of stable attractors of a CA are the same for both update regimes (Schönfisch and de Roos [1999]). Indeed, any configuration that is stable under one updating regime is stable under any other since it is of no consequence in what order the cells are updated. In addition, various patterns formed in synchronous updating are absent with asynchronous updating. For example, in the two-dimensional version of the Iterated Prisoner's Dilemma (see Section 5.3), Huberman and Glance [1993] found that the complex mosaics of defecting and cooperating cells attained with synchronous updating completely disappeared with asynchronous updating and the CA reached a fixed state of entirely defecting cells. Similarly, we have already mentioned that the Game of Life also converges to a stable state with asynchronous updating.

So, the question arises as to which updating regime is the most appropriate for modelling physical phenomena. Some argue that asynchronous updating is more natural since there is no universal clock in Nature. The highly geometric patterns produced in synchronous updating dissolve away with asynchronous regimes and hence are really artifacts of the updating regime. However, weighing in on the side of synchrony we have M. Sipper [1995], who states, "It may be argued that from a physical point of view synchrony is justified: since we model a continuous spatial and temporal world we must examine each spatial location at every time step, no matter how small we choose these (discrete) steps to be." However, as pointed out by Schönfisch and de Roos [1999], "the difference between synchronous and asynchronous update is a question of how we look at the (real) process. If we observe only in large time intervals we will see that all cells have been updated (at least) once in one time step, implying synchrony. If we refine the time scale such that in every time interval at the most one event will happen, then we find asynchrony... . In summary, it will depend on the actual cellular automata how strong the influence of different updating methods will be."

There are various methods to update cells asynchronously and these fall into two distinct categories, *step-driven* and *time-driven*. In step-driven updating, each cell of the array is updated by some algorithm one cell at a time. For example, in the simplest case, a fixed directional line-by-line sweep of each row can be made to update each of the cells sequentially. Or, the cells can be randomly ordered and each pass is made through this ordering or one can even take a different ordering for each pass. In time-driven updating, each cell has an internal clock that

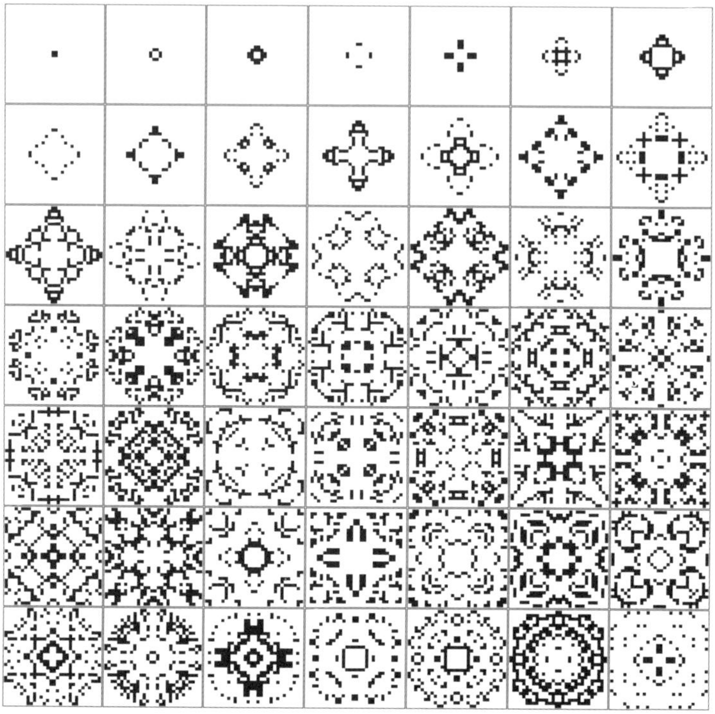

Figure 4.31 Starting with four black cells and the CA of Roli and Zambonelli in the text under synchronous updating. These are the first 49 configurations with many similar that follow.

"wakes up" the cell at some specific point in time so that its state can be updated. This waking up can also be done in a stochastic manner so that each cell will be updated with a certain probability, p_c.

In the sequel, we consider an example found in Roli and Zambonelli [2002] of the local rule:

- A dead (black) cell becomes alive (white) if it has two alive neighbors.
- A living cell remains alive if it has one or two neighbors alive, otherwise it dies.

These criteria are a watered-down version of the Game of Life, but it has some star-studded dynamics of its own in the synchronous case (Fig. 4.31).

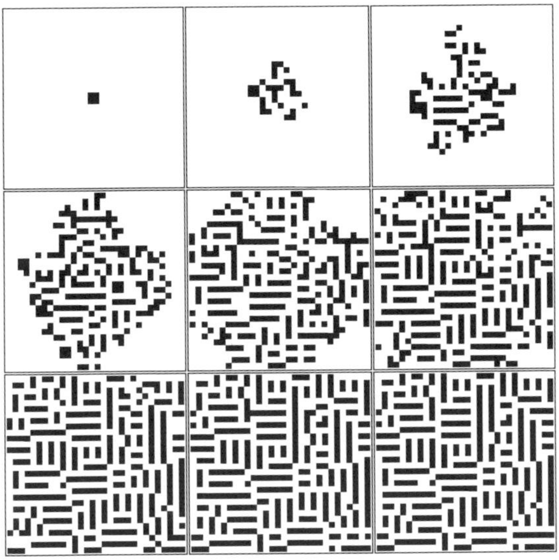

Figure 4.32 What a difference a delay makes! The asynchronous updating of the same CA with each cell being updated or not with a probability of $p_c = 0.5$ at each time step. The configurations are a sampling every 10 time steps and the final configuration is stable.

However, starting with the same block of four black cells in the asynchronous case is much less dramatic and leads to a stationary state as in Fig. 4.32.

Roli and Zambonelli considered a third type of updating regime in addition to synchronous and asynchronous that they termed *dissipative*. These CA (called *dissipative cellular automata — DCA*) have asynchronous time-driven updating but also allow an external perturbation to change the state of any of the cells of the array concurrently with the transition function (Fig. 4.33). This can be thought of in terms of the environment interacting with the automaton by providing "energy." The notion of a DCA was inspired by the dissipative systems much studied by Nobel laureate Ilya Prigogine and his school (cf., e.g., Nicolis and Prigogine [1989]).

One way to achieve interaction with the environment is for the perturbation to take the form of forcing cells at random to become alive with a certain probability, p_d. The degree of perturbation must be sufficiently high to affect the dynamics of the CA, but not too high so as to

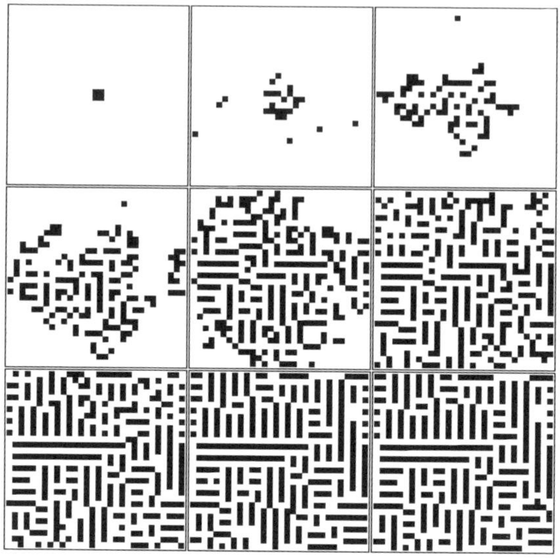

Figure 4.33 The evolution of the dissipative CA whereby each cell has a probability of $p_d = 0.001$ to be perturbed to be alive at each time step. Large-scale patterns evolve that are seen to persist but are not completely static. Actually, these patterns arise even without any black cells in the initial conditions due to the perturbations.

make the behavior essentially random. The ratio p_d/p_c is the crucial factor here, and although stationary configurations are not reached, large-scale patterns do persist.

Because of the open nature of DCA to the environment, the authors assert that "the dynamic behavior of DCA is likely to provide useful insight into the behavior of real-world open agent systems."

CHAPTER 5

APPLICATIONS

The important thing in science is not so much to obtain new facts as to discover new ways of thinking about them.

— *Sir William Bragg*

It becomes plausible that many, perhaps all, of the processes upon which life is based are algorithmically describable and that, therefore, life itself is achievable by machines.

— *Christopher Langton* – Artificial Life: An Overview

Cellular automata have applications to many diverse branches of science, such as biology, chemistry, physics, and astronomy. In the models that follow, we are able to accurately reproduce many physical phenomena when we attempt to emulate the underlying mechanisms in the setting of a cellular automaton. And this approach has some virtues. According to Ermentrout and Edelstein-Keshet [1993]: "CA models are fast and fairly easy to implement. Furthermore, the visual feedback they provide is striking and often resembles the patterns experimentally observed. These two aspects of CA modeling are its biggest advantages over more traditional approaches. CA are thus an excellent way of formalizing a theory of purported mechanism into computational terms. This is an important first step in the understanding of any physical process."

One of the traditional approaches to modeling natural phenomena has been through the use of differential equations. These equations describe how in some particular setting, a quantity such as, for example, temperature, changes in time and space. A rich and diverse literature for solving these equations has developed over the centuries, but of course there are certain inherent drawbacks with any attempt to model the physics of the real world. For a discussion of the differential equations vs. cellular automata approaches to modeling, see Toffoli [1984]. Various *Mathematica*-based CA models, including some that appear in this book, are found in Gaylord and Nishidate [1996], and Gaylord and Wellin [1994].

In ancient China, patterns appearing naturally were referred to as *li* and were thought to represent an underlying order in the world. Some of the ones described in the sequel can be found in the charming little book by David Wade [2003]. One of the first attempts to mathematically re-create forms found in Nature using simple rules was demonstrated by Alan Turing in a seminal work in 1952. Turing's inhomogeneous patterns were formed by the changing spatial concentrations of two chemical reactants in a solution and were described by a set of *reaction-diffusion* equations. These equations in two-dimensions describe how the two reactants having concentrations u and v respectively change with time:

$$\begin{aligned}\frac{\partial u}{\partial t} &= f(u,v) + d\,\Delta u,\\ \frac{\partial v}{\partial t} &= g(u,v) + d'\,\Delta v,\end{aligned}$$

where f and g are the reaction terms, and the diffusion terms are represented by the Laplacian operator $\Delta = \partial^2/\partial x^2 + \partial^2/\partial y^2$ times (diffusion) constants d, d'. This sort of interplay between two competing tendencies lies at the heart of various CA models, such as the first three models in the sequel.

5.1 EXCITABLE MEDIA

This term refers to a group of models that are characterized by a transition of a cell's state following a basic pattern: A quiescent state transits to an excited state, then to a refractory state before returning again to the quiescent state from which the cell can become excited again, and so forth. These models all tend to exhibit waves of various forms.

5.1.1 Neural Activity

Our first example is a cellular automaton model for generating waves, say to heuristically model a network of neuron excitation and recovery, and is due to Greenberg and Hastings [1978]. We have already seen a related model in Silverman's Brian's Brain. Either the von Neumann or Moore neighborhood (as in Brian's Brain) is considered. Each cell (or neuron) in a rectangular lattice has a quiescent state $= 0$, excited states $= 1, 2, .., e$ and refractory states $= e+1, e+2, ..., e+r$. Note that in Brian's Brain there was one excited ("firing") state and one refractory state. The G-H rules are as follows:

- A cell in state k, for $1 \leq k \leq e+r-1$, at the next time step becomes state $k+1$.
- A cell in the last refractory state $e+r$ at the next time step becomes state 0.
- A cell in the quiescent state 0:
  - at the next time step becomes the excited state 1 if one or more of its neighbors is in an excited state;
  - if none of its neighbors is in an excited state, then at the next time step its state remains 0.

Interpreting the quiescent state as "susceptible," the excited states as "infectious," and the refractory states as "immune," Birgitt Schönfisch [1995] has developed a similar epidemiological model whereby a susceptible individual becomes infectious if at least a threshold number of neighbors are infectious rather than simply if one or more neighbors are infectious. Schönfisch also considers a possibly more realistic scenario in a biological sense whereby a susceptible individual becomes infectious with a certain probability that increases with the number of neighbors that are infectious. These models are similar to the classical *SIR* model (**s**usceptible, **i**nfectious, **r**ecovered) for many infectious diseases. A deterministic example is depicted in Fig. C.6, see color section.

The first two conditions of the Greenberg-Hastings model can be considered as the reaction phase and the third condition is the diffusion phase. Simplified versions can have one excitatory state and/or one refractory state. In the simplest version we have quiescent $= 0$, firing $= 1$, refractory $= 2$, so that we can formulate the Greenberg-Hastings reaction (**R**)-diffusion (**D**) process as

$$c_{i,j}(t+1) = \mathbf{R}[c_{i,j}(t)] + \mathbf{D}[c_{i-1,j}(t), c_{i+1,j}(t), c_{i,j-1}(t), c_{i,j+1}(t)],$$

where

$$\mathbf{R} = \begin{cases} 2 & \text{if } c_{i,j}(t) = 1, \\ 0 & \text{otherwise}, \end{cases} \qquad \mathbf{D} = \begin{cases} k & \text{if } c_{i,j}(t) = 0, \\ 0 & \text{otherwise}, \end{cases}$$

and $k = 1$ if at least one of the von Neumann neighbors is firing and $k = 0$ otherwise.

As one would expect, varying the parameters leads to different wave forms propagating outward.

A reversible G-H model can be constructed by using Fredkin's idea of subtracting the state of the central cell at the previous time step (Tamayo and Hartman [1989]):

$$c_{i,j}(t+1) = \mathbf{R}[c_{i,j}(t)] + \mathbf{D}[c_{i-1,j}(t), c_{i+1,j}(t), c_{i,j-1}(t), c_{i,j+1}(t)] - c_{i,j}(t-1),$$

where the sum is taken $\bmod 3$ and the reaction-diffusion parts are the same as in the preceding. Here the initial conditions are quite significant as they are preserved in the subsequent dynamics. The authors have identified three basic regimes that the system eventually falls into:

- Regular regime (Fig. 5.1). Here regular wavefronts move about in an unperturbed environment.
- Random regime (Fig. 5.2). Regularity gradually breaks down and gives way to randomness, lacking any structure.
- Turbulent regime (Fig. 5.3). Ringlike structures emerge and form around source and sink regions that gradually become diffused.

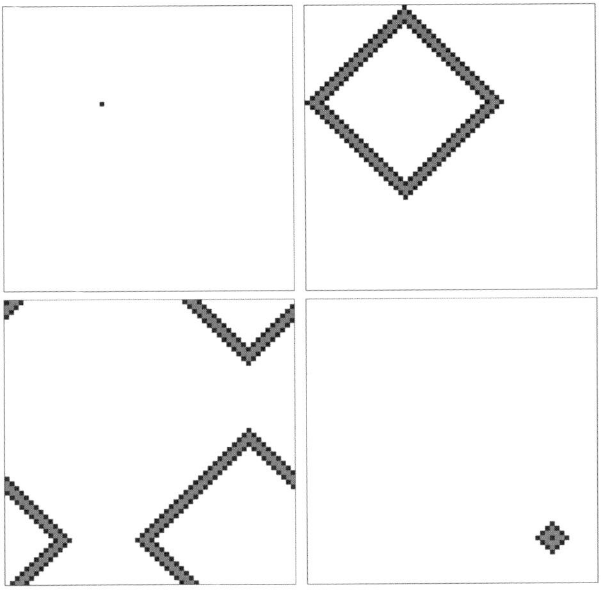

Figure 5.1 This is the regular regime at time steps 0, 20, 40, 60 (left to right, respectively). The initial conditions are: cell $(20, 20)$ is firing and all others are quiescent. The array is 60×60 with periodic boundary conditions.

Figure 5.2 This is the random regime at time steps 30, 240, 480, 840 (left to right, respectively). The initial conditions are: cell $(20, 20)$ is firing, $(21, 20)$ is refractory, and all others are quiescent.

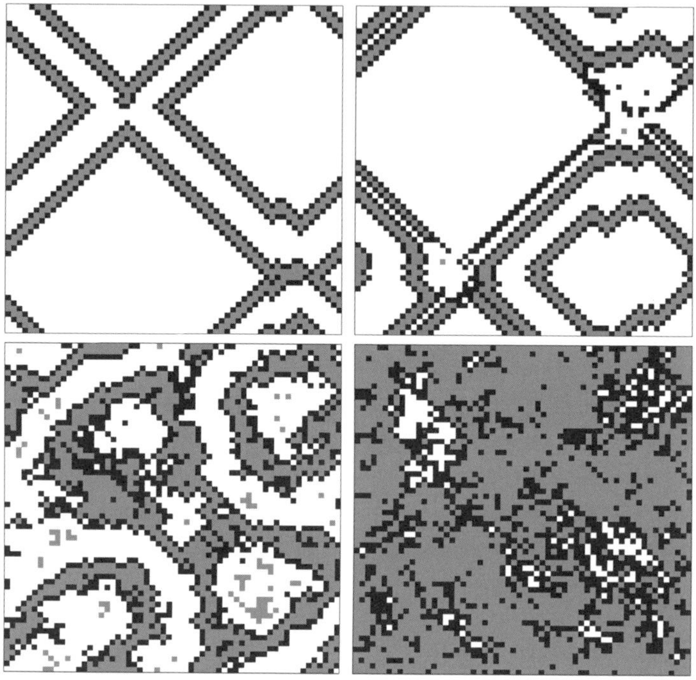

Figure 5.3 This is the turbulent regime at time steps 225, 450, 675, 900 (left to right, respectively). The initial conditions are: cells $(18, 18)$, $(23, 19)$, $(19, 22)$ are firing, and all others are quiescent. Wavefronts first interact (top right) and then form around source and sink regions (bottom left) that finally become diffused.

5.1.2 Cyclic Space

Another interesting wave generating model similar to the preceding one was created by David Griffeath of the University of Wisconsin and dubbed the Cyclic Space cellular automaton by A. K. Dewdney in his *Scientific American* column of August 1989. Again we employ a rectangular grid with each cell taking state values $0, 1, 2, ...,$ up to a maximum value of m and we can use either the four-cell von Neumann or the eight-cell Moore neighborhood. The array is then initialized with all cells taking on a random value from 0 to m. The rules for the automaton are:

- If a cell in state k, for $0 \leq k < m$, has any neighbor with state value $k + 1$, then at the next time step the cell takes on the value $k + 1$.

- If a cell has the maximum state value m and if any neighbor has the state value 0, then the cell at the next time step takes on the value 0.

- The cell state remains unchanged if there are no neighboring cells meeting the required conditions.

Note the similarities with the Greenberg-Hastings model. The second condition here is like the neuron cell state changing from the rested to the excited state. However, a point of difference is that in the neuron model a cell (in the excited and refractory states) updates its state value independent of the state of its neighbors, whereas in the cyclic space model the update depends on at least one neighbor being in the next higher state. We can also think of a cell in state k having a neighbor in state $k + 1$ as being "eaten" by the latter.

According to Griffeath, there are four phases that the automaton goes through: debris, droplet, defect, and demon, which are illustrated in the Fig. C.7. Starting from an initially random configuration, which soon looks like debris, we find that droplets of solid color begin to emerge and grow by eating neighbors. Then certain nonintersecting loops of cells can form "defects" that initiate the next phase of spiral formation. A defect occurs when each of the cell states of the loop differs by $+1, -1$, or 0 from the next cell in the loop (bearing in mind that the highest state differes from the lowest by 1), and furthermore when all of these flux changes are added consecutively, the sum is nonzero. The newly formed spirals grow larger with some being absorbed by others, leading to the final stage of demons that are spiral configurations that continue to cycle in their own confined region and occupy the entire space (Fig. C.7, see color section).

The first three phases are what are called *metastable* insofar as they persist for many generations. The reason for this persistence is because it is not very likely that a cell will change its state value. In fact, using a von Neumann neighborhood and n state values, the probability p that the central cell will change its value is given by 1 minus the probability that none of its four neighbors is one integer value higher, that is,

$$p = 1 - \left(\frac{n-1}{n}\right)^4,$$

since the probability that a single neighbor is not one integer value higher is $(n-1)/n$. For $n = 14$, we find that $p = 0.256$, or about one chance in four. And this is the case when all the neighbors can have random values. If all the neighbors have the same value as the central cell, then, of course, the cell remains constant until the situation in the neighborhood changes. The preceding formula also says that the higher the number of states, the longer the phases persist.

5.1.3 The Hodgepodge Machine

Certain types of complex chemical reactions oscillate displaying features of waves and spirals. One such reaction is the combination of carbon monoxide (CO) with oxygen to form carbon dioxide (CO_2) catalyzed by tiny palladium crystallites. Because of the multitude of the latter, a cellular automata model was suggested by Martin Gerhardt and Heike Schuster [1989] of the University of Bielefeld whereby each cell represents a catalytic unit.

For each cell, there are $n+1$ states given by $0, 1, 2, ..., n$. According to Gerhardt and Schuster, a cell in state 0 is "healthy," a cell in state n is "ill," and all states in between $1, 2, ..., n-1$ are "infected." Infection represents the level of oxygen saturation of each catalytic unit and increases until a maximum saturation level is reached (ill). There are three rules governing the transition to the next state that reflect the chemistry of the reaction:

- If $c(t) = 0$ (cell is *healthy*, no oxygen), then it will become infected if a suitable number of neighboring cells are infected:

  $$c(t+1) = \left[\frac{N_{\text{ftd}}}{k_1} + \frac{N_{\text{ill}}}{k_2}\right],$$

  where N_{ftd} is the number of infected cells in a neighborhood (either von Neumann five-cell or Moore nine-cell), N_{ill} is the number of ill cells in a similar neighborhood, and k_1, k_2 are constants representing minimal levels for the cell to become infected. That is to say, if the expression in the brackets is less than 1 because N_{ftd} and N_{ill} are not sufficiently large enough, then $c\,(t+1)$ remains 0 and thus healthy. Here the square brackets means to round down to the nearest integer.

- If $1 < c\,(t) < n$ (cell is *infected*), then the degree of infection increases at the next time step to

$$c\,(t+1) = g + \left[\frac{\Sigma}{N_{\text{ftd}}}\right],$$

  where g is a (constant positive integer) "infection rate" and Σ is the sum of the infected state values of the cell neighborhood. If the value computed on the right-hand side of the expression exceeds n, then the value for $c\,(t+1)$ is taken to be just n. Here we see that the infection state of a cell increases by the fixed amount g plus a local average of infection.

- Finally, if $c\,(t) = n$ (cell is *ill*), then $c\,(t+1) = 0$, that is, ill cells become healthy due to a phase transition of the catalytic units.

To run the colorfully named "hodgepodge machine," one must first specify the number of cell states n, the constants k_1, k_2, and the infection rate g. A typical example has values $n = 100$, $k_1 = 2$, $k_2 = 3$, and $g = 20$. One starts with a random initial configuration and after awhile the system has the appearance (although constantly changing) of Fig. C.8, see color section.

This pattern appears to emulate the actual wave pattern depicted in the well-known Belousov-Zhabotinskii reaction: the oxidation of malonic acid by potassium bromate with iron or cerium as a catalyst (Fig. 5.4). The chemical waves of this reaction are normally described in terms of partial differential equations in the form of reaction-diffusion equations; one popular model is known as the Oregonator, whose name betrays its origins. Another application is to the chemical waves and spiral patterns seen in the slime mold formation (Chapter 6).

Various investigators have developed other excitable cellular automata models that have, for example, multiple-armed spirals (Madore and Freeman [1983]) or three-dimensional twisted scrolls (Panfilov and Winfree [1985]).

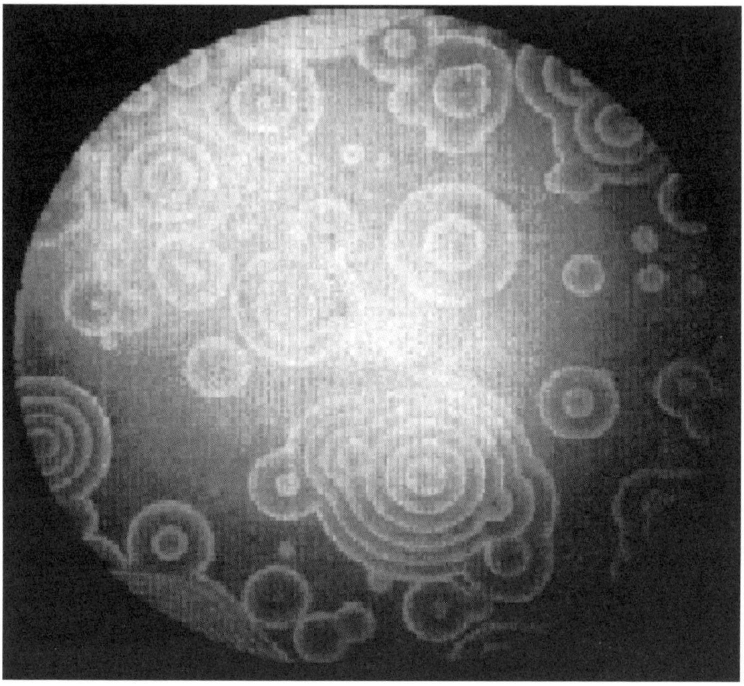

Figure 5.4 Wave patterns seen on the surface in the Belousov-Zhabotinskii reaction.

5.2 SCHELLING SEGREGATION MODEL

Since human interactions can and often do function on a local level, it is not surprising that cellular automata have also found various applications in the social sciences. The earliest known work seems to have been James Sakoda's "The checkerboard model of social interaction" [1971], although the model had its origins in Sakoda's unpublished 1949 dissertation, which takes us back to the very origins of the Ulam/von Neumann cellular automata theory itself. A now classical model due to Harvard economist Thomas Schelling in 1971 has to do with the formation of segregated neighborhoods. He believed that people had a certain tolerance level for different ethnic groups and felt comfortable with where they lived as long as a certain proportion of individuals of their own kind lived in their neighborhood. In general, when the neighborhood fell below this proportion, then the individual would move to another neighborhood more to their liking. Schelling used two types of coins representing two different ethnic groups on graph paper which he distributed over some of the grid sites (leaving some blank regions) and which obeyed the following simple contentment criterion:

- If the proportion of neighbors in an eight-cell Moore neighborhood that are of the same kind (state) as the central cell becomes lower than some threshold x, then the central cell moves to a "nearby" site in which the neighborhood has a proportion of neighbors of the same kind that is at least x.

Schelling was interested in finding out what values of the threshold x produced segregated neighborhoods and it was quite surprising to find that, even with quite low values for x, initially integrated neighborhoods resulted in high levels of segregation. Of course, in reality, not all people living in a neighborhood would have the same contentment threshold and many would not care who the neighbors were as long as they were quiet. Moreover, as Schelling pointed out, there are other mainly economic factors that have a segregating effect. So the model is mainly of heuristic value but does indicate an emergent feature that is rather unexpected. However, it is pleasing to note that Schelling was a joint winner of the 2005 Nobel Prize in Economics in part for this work.

To implement the Schelling model, by nearby we take Moore neighborhoods of ever-increasing radius until a suitable neighborhood is found or until all sites have been tried. The updating of cell sites is done asynchronously by choosing a cell site at random and, if it is occupied, then applying the above contentment criterion. This is done until the system reaches an equilibrium state. Schelling found that 25% to 30% of vacant space was the most desirable in the initial random configuration of equal numbers of black and white cells. On average it is found that even a low contentment threshold of 30% leads to 70% of same cell state neighbors (see Fig. C.9, see color section).

It is worth noting that neither Sakoda nor Schelling used the term *cellular automaton* at any point in their work. Moreover, it is interesting that there is now even a *Journal of Artificial Societies and Social Simulation,* no doubt a direct descendent of the pioneering works of Sakoda and Schelling. A *Mathematica*-based toolkit for the CA modeling of social interactions can be found in the book by Gaylord and D'Andra [1998] and the Schelling model is also discussed in Gilbert and Troitzsch [1999]. Hegselmann and Flache [1998] even plead for more CA based modeling of social dynamics.

5.3 PRISONER'S DILEMMA

The Prisoner's Dilemma is a classic model of the dynamics of cooperation and noncooperation, and made its first appearance in a lecture given in 1950 by mathematician Albert W. Tucker. In order to play a single round of the Prisoner's Dilemma, two players, A and B, have but two options and must decide whether on not they will cooperate (C) or defect (D). Typical payoffs to each player are indicated by the following table:

		B	
		Cooperate (C)	Defect (D)
A	Cooperate (C)	3\3	0\5
	Defect (D)	5\0	1\1

Thus, if both players decide to cooperate, they are rewarded with three points each, whereas if both players defect, they obtain only one point each. What is known as the "sucker's payoff" is when player A defects and receives five points but player B cooperates and ends up with zero points.

What does all this have to do with prisoners? It is based on the scenario of two people (A and B) being arrested for committing some crime (and incidently, was featured in the David Bowie movie *The Labyrinth*). Both A and B are interrogated separately by the police, who do have some evidence against them. We will use the value of $5 - p$, where p is the number of points obtained by one of the apprehended, to calculate their prison terms. So, if they both deny their guilt (that is, they cooperate with each other), they get relatively light two year sentences since they were each rewarded with three points. However, if they both confess to the crime (i.e., defect), then they each get relatively severe four year sentences reflected in their mutually low reward of one point. In the sucker's payoff, one player makes a deal with the police admiting guilt (defects) while the other one still denies it (cooperates), with the latter getting the harshest prison sentence of five years (since their reward was zero) and the former getting none (with reward five).

If you are one of the players, what should your response be? Clearly, no matter what the other player does, you should defect. If they happen to cooperate, then you've scored a massive five points to their nil and they are a sucker for it. Even if they have defected too, you would still score something (one point) whereas if you had cooperated you would end up with nil yourself. Of course your partner also faces the same situation and should choose defection as well. This *defect-defect* optimum strategy is known as the *Nash equilibrium* (due to Nobel Prize winner John Forbes Nash) of the Prisoner's Dilemma game.

Viewed collectively, it is certainly true that if you both cooperate then the rewards (three points each) are greater than if you both defect (one point each), but from an individual's perspective, there is the temptation of even greater rewards from defection if only the other cooperates, leaving them with nothing. Thus the potential personal gain from defection outweighs the potential collective gain from cooperation.

The Prisoner's Dilemma is but a single example from the vast field of *game theory*, the formal study of which was begun by John von Neumann in the late 1920s, although some earlier work had been done by others. It was introduced to the world on a grand scale in the 1944 classic text by John von Neumann and Oskar Morgenstern, *Theory of Games and Economic Behavior.* Game theory became serious business during the Cold War period and has since found many important applications in the study of evolutionary biology, political science, psychology, sociology, and economics. It has also become a course taught in its own right at universities. For a delightful discourse on the Prisoner's Dilemma, John von Neumann, and game theory, see Poundstone [1993].

It becomes far more interesting, however, if the game is played over and over in what is known as the Iterated Prisoner's Dilemma (dating back to M. M. Flood [1952]). The object for each player is to accrue the largest points total which now requires that some overall strategy be adopted. According to logician Patrick Grim, "It is no simplification to say that our strongest and simplest models of the evolution of biological and sociological cooperation — and in that respect our strongest and simplest models of important aspects of ourselves as biological and social organisms — are written in terms of the Iterated Prisoner's Dilemma."

One simple strategy would be to always cooperate, but this approach is easily taken advantage of. And always defecting can lead to low total scores. Political scientist Robert Axelrod of the University of Michigan conducted a round-robin tournament between strategies submitted by colleagues (see the famous paper by Axelrod and evolutionary biologist W. D. Hamilton [1981] in *Science*). Each game between competing strategies consisted of 200 moves, yielding possible scores from 0 to 1000, so that a "good performance" score was judged to be 600 points (complete mutual cooperation) and a "poor performance" score would be 200 points (complete mutual defection). In the first round of the tournament, the overall winner with an average of 504 points per game, submitted by game theorist Anatol Rapoport, was called Tit-for-Tat. In this simplest of strategies, a player cooperates on the first move and then does exactly what the other player did on the previous move. While never actually attaining a higher score in any particular game, this strategy can lead to many rounds of cooperation between opponents with its commensurate rewards, whereas the more defecting strategies do poorly. While Axelrod has shown that there is no best strategy over all competing environments, he has shown that Tit-for-Tat is very robust over a range of environments. In fact, Tit-for-Tat has many strengths: "What accounts for Tit-for-Tat's robust success is its combination of being nice [not being the first to defect], retaliatory, forgiving, and clear. Its niceness prevents it from getting into unnecessary trouble. Its retaliation discourages the other side from persisting whenever defection is tried. Its forgiveness helps restore mutual cooperation. And its clarity makes it intelligible to the other player, thereby eliciting long-term cooperation" (Axelrod [1984]).

On the other hand, it was pointed out by Axelrod that at least three other strategies would have won the first round over Tit-for-Tat. One of them was the example strategy he sent to prospective contestants, called Tit-for-Two-Tats but inexplicably no one submitted it! This strategy defects only when the opponent has defected on the *two* previous moves and is thus more forgiving than Tit-for-Tat. In a second round of the tournament conducted by Axelrod, with a much larger pool of competitors, Tit-for-Tat won again in spite of everyone knowing the results of the first round. Tit-for-Two-Tats was submitted this time but did quite poorly as some competing strategies were able to capitalize on its tolerance of a single defection. According to Axelrod, "The moral of the story is that the precise level of forgiveness that is optimal depends upon the environment."

Axelrod [1987] also performed computer experiments in order to evolve genetic algorithm strategies to play the Iterated Prisoner's Dilemma. The opposition was eight human developed strategies and it was found that most of the evolved GAs had similar traits to Tit-for-Tat and some of them performed even better *in that particular environment.*

A spatial component can be added to the game so that it is played on a square grid against multiple partners. One can take those partners to be the cells of a von Neumann or Moore neighborhood with the central cell playing off against each one (employing periodic boundary conditions). In one adaptation each player will employ the strategy of either simply cooperating or defecting. At each time step, every square plays one round of the Prisoner's Dilemma against its four or eight neighbors and the score for each cell is then totaled. Each cell at the next time step will then adopt the strategy of the most successful cell in its own neighborhood. Clearly a cooperator surrounded completely by defectors will get swallowed up and at the next time step become a defector. The outcome over time very much depends on the payoff values and often a stable equilibrium is reached. A rigorous treatment of the evolution of cooperation in the spatial prisoner's dilemma setting was made by Schweitzer, Behera, and Mühlenbein [2002] utilizing a von Neumann neighborhood, and asynchronous as well as synchronous interactions with long term memory have been considered by O. Kirchkamp [2000].

There is another way to depict the payoff values. Let R be the reward payoff for mutual cooperation, T the temptation payoff for defecting in the hope your opponent cooperates, S the sucker payoff for cooperating when your opponent has defected, and P the punishment payoff for mutual defection.

		B	
		Cooperate (C)	Defect (D)
A	Cooperate (C)	R\R	S\T
	Defect (D)	T\S	P\P

In the Prisoner's Dilemma discussed above, $\{R; S; T; P\} = \{3; 0; 5; 1\}$, but more generally it can be assumed that $T > R > P > S$ and $R > (S + T)/2$. This latter condition ensures that the reward for mutual cooperation should still be greater than that earned if players in turn exploited each other (Fig. C.10, see color section).

Nowak and May [1992] used the nonstandard payoff values $\{R; S; T; P\} = \{1; 0; b > 1; 0\}$ (note that $S = P$ so that this is not strictly a Prisoner's Dilemma), which permitted a measure of control over the outcome by adjusting the temptation parameter b. In this scenario, in a square block of four cooperators surrounded by a sea of defectors, the cooperators will score a total of four points (if we also include cells playing against themselves), whereas any nearby defectors are only exposed to two cooperators, resulting in their scoring $2b$ points. Thus if b is less than two, clusters of cooperators can hold their own against hordes of defectors (Fig. C.11, see color section).

Starting with a single defector cell, a mosaic of regions of defection and cooperation can evolve that allows them to coexist indefinitely in an ever-changing state of flux (Fig. C.12, see color section).

Another approach to the spatial iterated prisoner's dilemma is to allow for various strategies to compete against each other. Initially, a particular strategy can be randomly assigned to each cell of the array. Each cell then competes for a fixed number of rounds with each of its eight neighbors using the standard payoffs. At the end of each session, a cell will adopt the strategy of its most successful neighbor in terms of highest total score. If two or more neighbors have the highest score, then the central cell will retain its present strategy. In Fig. C.13 (see color section) we have depicted just such an event with five competing strategies: Always Cooperate, Tit-for-Tat, Random, Pavlov, and Always Defect.

The Random strategy simply cooperates or defects according to the head or tails outcome of a virtual coin toss. It came last place in Axelrod's first round. The other new strategy, Pavlov, repeats its former choice whenever it earns a high payoff like R or T and switches that choice whenever it earns a low payoff like P or S. Here the philosophy is: If winning, hold fast; if not, change course. With time, Tit-for-Tat is seen to predominate.

There could well be biological payoffs for all this. According to Nowak, May, and Sigmund [1995],

> ... in simpler organisms, cooperation persists, perhaps by virtue of self-organized spatial structures generated by interactions with immediate neighbors in some fixed spatial array. In the course of evolution, there appears to have been ample opportunity for cooperation to have assisted everything from humans to molecules. In a sense, cooperation could be older than life itself.

5.4 BIOLOGICAL MODELS AND ARTIFICIAL LIFE

Cellular automata models have proven very useful in biology for as Ermentrout and Edelstein-Keshet [1993] put it, in biology, "unlike physics, there are few 'laws' such as the Navier-Stokes equations or Newton's laws." So one has to create models of systems by extracting the most salient features of the interactions of various components over time and describing these in some suitable framework, be it differential equations or cellular automata. A very early model for filamentous cell growth using local interactions between cells to update their state in discrete time was proposed by A. Lindenmayer already back in 1968. Detailed and useful CA models have arisen in the work of, say, Kaplan *et al.* [1988] on the dynamics of cardiac conduction, which was studied using an excitable CA model in a spatially inhomogeneous medium, or Pytte *et al.* [1991], who deduced an elegant model of a region of the hippocampus (a structure in the floor of the brain) with up to 10,000 neurons which gives qualitative and quantitative data that otherwise would require the numerical solution of about 250,000 coupled differential equations. CA models have been made of the immune system (De Boer and Hogeweg [1992], Celada and Seiden [1992]), tumor growth (Moreira and Deutsch [2002]), and genetic disorders (Moore and Hahn [2000, 2001]) among numerous others. An informative discussion of cellular automata models in biological modeling can be found in the survey paper by Ermentrout and Edelstein-Keshet [1993] and in the text by Deutsch and Dormann [2005].

Biological CA models can also be considered as a form of *artifical life (AL* or *Alife)*, which, according to Chrisopher Langton, one of the subject's founders, is "devoted to understanding life by attempting to abstract the fundamental dynamical principles underlying biological phenomena, and recreating these dynamics in other physical media, such as computers, making them accessible to new kinds of experimental manipulation and testing." So much work now goes on in this subject that it has its own journal: *Journal of Artificial Life*; see also the text of Adami [1998].

Of course the burgeoning field of AL goes beyond the scope of cellular automata, as, for example, in the *Blind Watchmaker* program of Richard Dawkins [1989] that generates a whole menagerie of two-dimensional artificial organisms called "biomorphs" that are found to emerge under suitable biologically inspired conditions or the highly addictive *Creatures* program created by Steve Grand that utilizes a combination of genetic-based algorithms and artificial neural networks. The *Avida* AL research program can be downloaded at `http://dllab.caltech.edu/avida/` and creates an environment in which digital organisms can replicate, mutate, and evolve under conditions simulating natural selection. The *Avida* program itself evolved from Alife pioneer Tom Ray's ground-breaking *Tierra* program (`http://www.his.atr.jp/~ray/tierra/source/index.html`). As well, there is the *Terrarium* Web-based ecosystem game in which players create their own creatures that have specifically programmed attributes and which then compete for survival against the creatures developed by others. And we should not forget that Mary Shelley's Dr. Victor Frankenstein knew a thing or two about Alife. A comprehensive collection of artificial life resources can be found on the website `http://zooland.alife.org/index.html`, maintained by Jörg Heitkötter.

Further biological/artificial life models are also considered in the next chapter on complexity.

5.4.1 Genetic Algorithms

One of the remarkable things about biological organisms is that they have evolved to solve very complex tasks — the swimming skill of a shark, the flying skill of a bat, the three-dimensional vision of the reader. All of these have evolved through a gradual process of natural selection, so it is tempting to apply this wonderful mechanism of Nature to the computational solution of various problems. This was first discussed in the 1950s (Box [1957]) and subsequently explored by others, with the modern theory of what are now known as *genetic algorithms* (*GAs*) being developed by John Holland with his students and colleagues at the University of Michigan in a series of works in the 1960s and 70s (see Holland [1966, 1988, 1992]). See also the monograph by Mitchell [1996] for an excellent introduction to this fascinating subject. The wide-ranging list of applications of GAs includes population genetics, immunology, ecology, economics, machine and robot learning, and automatic programming.

So how do we artificially mimic natural selection in the form of a genetic algorithm? For a given task to be solved, we begin with a random population of algorithms that act as first approximations to

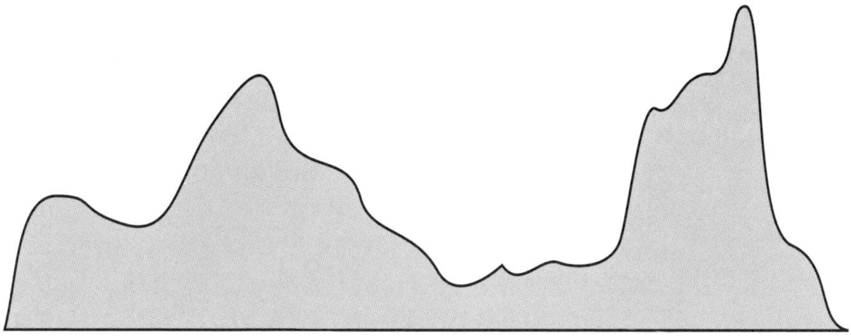

Figure 5.5 Graphical representation of a fitness landscape illustrating the peaks toward which the most fit solutions of a GA migrate and the valleys in which the least fit lie. Note that there can be peaks of varying heights.

solving the problem. These are encoded in some way, typically as strings of binary digits. After each algorithm has been run, they are all evaluated by a "fitness test," that is to say, as to how close they come to actually solving the problem. The most fit algorithms are then combined and mutated in order to form a new population of algorithms which are again subjected to the fitness test, and so on. After several "generations," one expects to have created a number of highly fit algorithms for the given task.

To visualize all this, the fitness function acting on the space of all possible solutions generates a graph of peaks and valleys called the "fitness landscape." The most fit solutions are near the peaks and the least fit lie in the valleys and typically an initial population will climb one of the peaks (a local maximum) as it evolves and becomes more fit (Fig. 5.5).

There are three basic genetic operations that can be performed on the algorithms:

- Exact reproduction whereby multiple copies of an algorithm are taken (*cloning* or *reproduction*).
- Swapping some part of their genetic encoding with another algorithm (called *crossover* or *recombination*).
- Random *mutation* of some part of their genetic encoding.

As an illustration of how these operations come into play, we present a genetic algorithm that was created by Auckland University student Michael Brough. Starting from an array in an initial random configuration of black ($= 1$) and white ($= 0$) cells, our aim is to turn the whole array into a checkerboard pattern. (Note that this is not completely possible since initial conditions having all cells in the same state will remain so, but this is of no consequence.) A nine-cell Moore neighborhood will be used along with periodic boundary conditions. Since each cell can be in either one of two states, there are $2^9 = 512$ different neighborhood-states. Any particular transition function will tell us the state value of the central cell (either 0 or 1) at the next time step for each of the 512 neighborhood-states and so can be represented by a binary string 512 digits long, such as 1001101...110. This is analogous to representing Rule 150 as the rulestring 10010110, only now our strings are somewhat longer. In keeping with a biological milieu, the rulestrings are thought of as "chromosomes" and the bits in the rulestrings are "genes." For our example, a hundred or so chromosomes are chosen at random to get things underway.

First of all, each chromosome is given a fitness test in the form of how well it converts an array of random black and white cells into an approximation of a checkerboard. This test is repeated on a few random initial configurations to give a fair evaluation of each chromosome's performance. Choosing the fitness function must be done with considerable care since one that is a bit flawed will not deliver the end result. In our particular example, points are awarded to a chromosome for producing patches of cells in a checkerboard pattern (after 100 time steps) according to the scheme:

For each cell (x, y):

- Subtract 3 points if cell $(x + 1, y)$ or $(x, y + 1)$ is in the same state as (x, y).
- Otherwise, add 8 points if cell $(x + 1, y + 1)$ is in the same state as (x, y) and subtract 5 points if it is not, and likewise add 8 points if cell $(x + 1, y - 1)$ is in the same state as cell (x, y) and subtract 5 points if it is not.

Next, all chromosomes are ranked from highest to lowest according to their fitness score. Each chromosome is then copied (*cloned*) a certain number of times that depends on its level of fitness compared to the average fitness score for all the chromosomes. In this way, those chromosomes that are below average are filtered out ("survival of the fittest"). Of course, there are other schemes that can be implemented here that pass on all chromosomes to the next phase, but the idea is to weight the chromosomes in some fashion toward the most fit.

The remaining chromosomes are allowed to "breed" with each other, that is, exchange some of their genetic material (0's and 1's). In our example, pairs of chromosomes are chosen at random from the initial population and these now become "parents." Typically, in other scenarios, designated blocks of digits are exchanged between parents (*crossover*), but in this instance adjacent digits have no particular relation to one another. So what has been done instead is that a new "baby" string is created by selecting one of the state values at each of the 512 digit positions randomly from either one of the parents. This new baby string is then added to the population of chromosomes. This procedure resulting in new offspring is a digital version of actual sexual reproduction.

Finally, a few bits in each chromosome are adulterated (flipping 1 to 0 or 0 to 1) by random *mutation* — one or two bits worked best. There needs to be some mutation in order to introduce new genetic material, but if there is too much, then the positive qualities found at each generation can be lost.

The resulting new set of chromosomes is now given the fitness test again and the above procedure repeated until the desired task is achieved or not. The proof is in the evolution (Fig. 5.6).

With the above genetic algorithm, it was found that it was successful most of the time and almost worked the rest of the time (producing patterns that were mostly correct but having minor flaws in them). Note that the graph of the fitness (Fig. 5.7) shows a gradual increase (although at the beginning sometimes there are decreases between generations demonstrating a random tendency at first), then a rapid jump, followed by further gradual increases until the maximum possible fitness score is attained. The GA with this maximum fitness score may not be successful on all possible initial configurations but has been so on all the ones given to it. At this stage one could start making the fitness function more stringent.

Figure 5.6 The evolution of the checkerboard pattern produced by the GA described in the text at 0 (random configuration), 10, 30, 50, 80, and 100 time steps.

Thus we have employed the essence of natural selection to solve a problem in pattern formation, an example of *emergent computation*. An evolved one-dimensional CA to produce globally synchronized alternating rows of completely black and white cells from arbitrary initial conditions was developed by Das, Crutchfield, Mitchell, and Hanson [1995]. Further examples in AL are discussed in Chapter 5, and for more on GAs, see Mitchell and Forrest [1994]. The evolutionary paradigm has also spawned related fields such as evolutionary programming, evolutionary strategies, and genetic programming.

Genetic algorithms fall under the purview of *complex adaptive systems*, that is, self-organizing systems of autonomous agents that over time exhibit some form of adaptive behavior. A beehive, a slime mold, and an ant colony are among the many complex adaptive systems found in Nature (see Chapter 6) and this notion has even found its way into the world of business and commerce. A very readable account of this whole circle of ideas can be found in the book by Nobel Prize winner Murray Gell-Mann [1994].

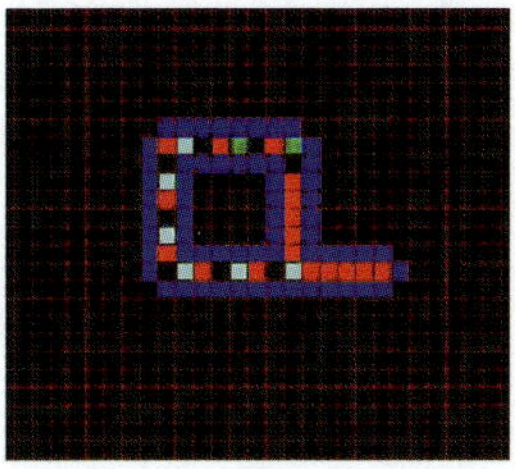

Figure C.1 The Langton looped pathway in its initial configuration. The different colors represent the different cell states.

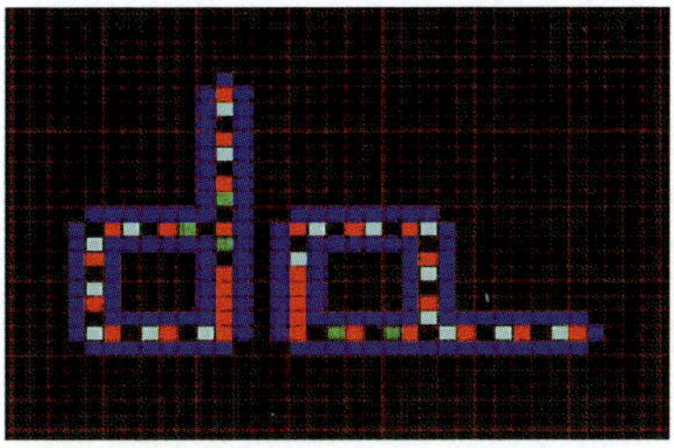

Figure C.2 The parent (left) and offspring loops each propagating new construction arms after their connection has been severed.

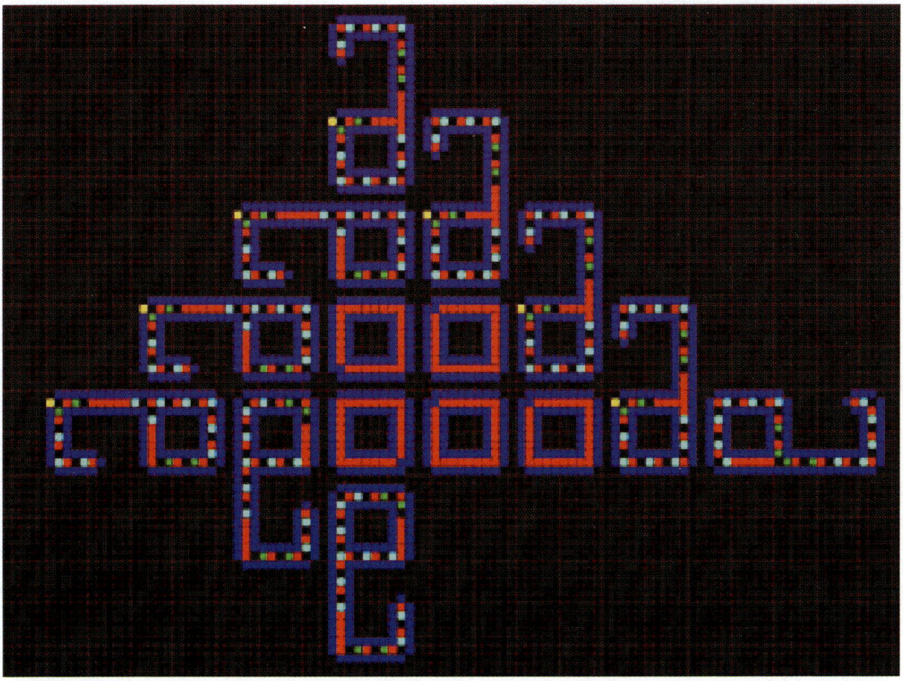

Figure C.3 The replication continues indefinitely filling up the plane with loops.

Figure C.4 The minimal time solution of Jacques Mazoyer of the firing squad synchronization problem using 6 states. Here $n = 34$ and time increases from top to bottom. The general is on the right-hand side and also reaches the firing state (black) together with all the soldiers in 67 time steps.

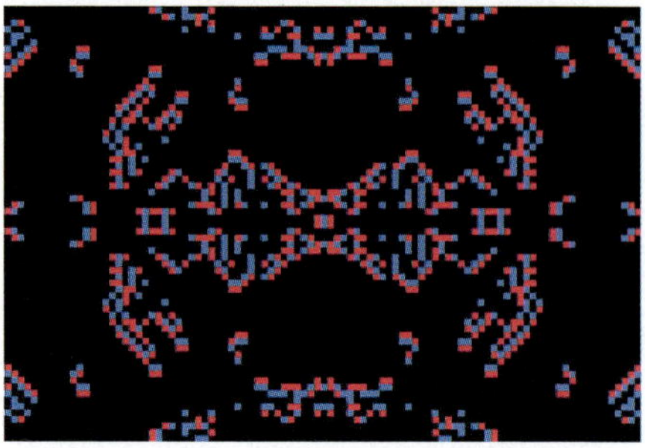

Figure C.5 The three-state cellular automaton Brian's Brain created by Brian Silverman. Waves of live cells tend to sweep across the array. Here the colors are: black = ready, red = firing, blue = refractory.

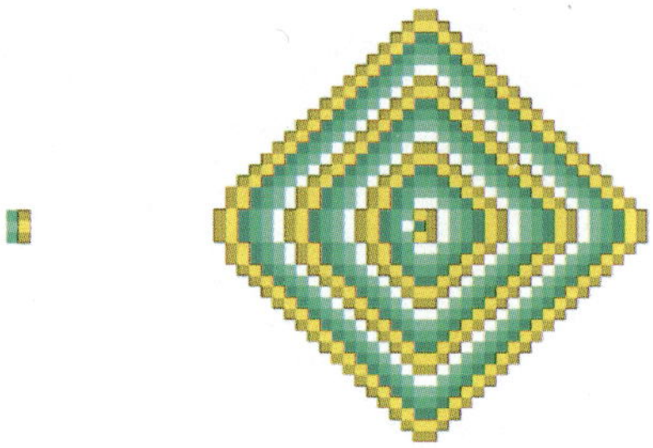

Figure C.6 In this Greenberg-Hasting model using the Moore neighborhood, there are 2 excited stages (yellow) and 3 refractory stages (green). As above the transition at each stage is from darkest to lightest. White is the rested state. At least 2 excited neighbors are required for a cell in the rested state to become excited. The figure shows the initial configuration (left) and after 19 iterations (right). Waves are generated in the center of the figure and grow and move outward.

Figure C.7 The time evolution from a random initial state to the formation of waves of the cyclic space automaton. The frames (left to right) represent the debris, droplet, defect, and demon phases, respectively.

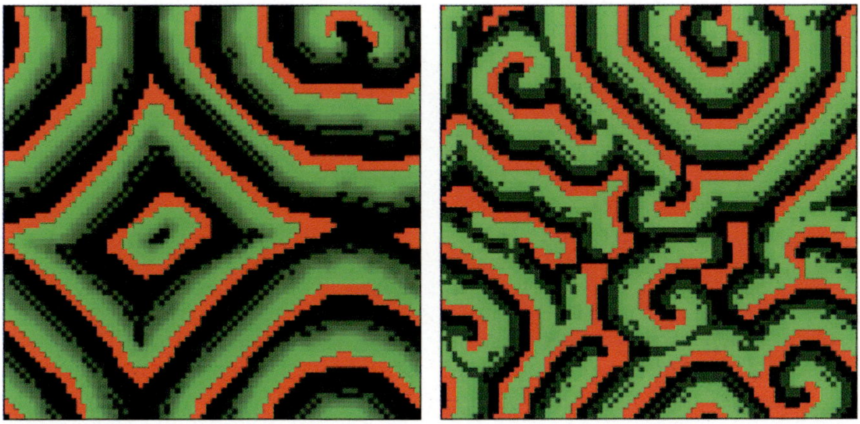

Figure C.8 Output from the hodgepodge machine using $g = 30$ (left) and $g = 40$ (right), with the other parameters given in the text. Notice the wave and spiral formations.

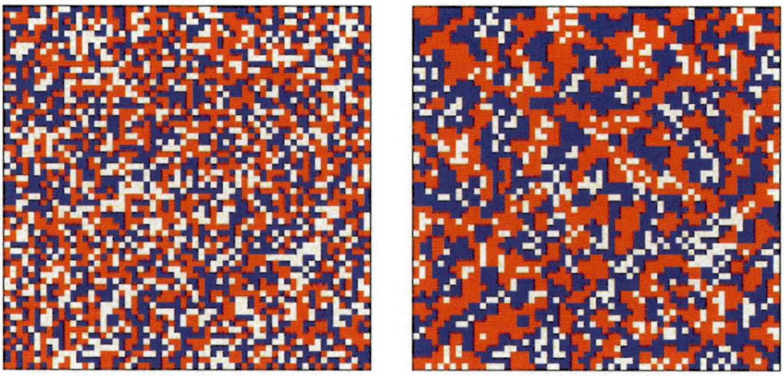

Figure C.9 In this Schelling model implementation, the left-hand side is an initial random distribution of occupied cell sites (25% white vacant space) and the right-hand side represents the segregated equilibrium state of the red and blue cells reached after applying the contentment requirement that 37.5% of the neighbors should be of the same kind.

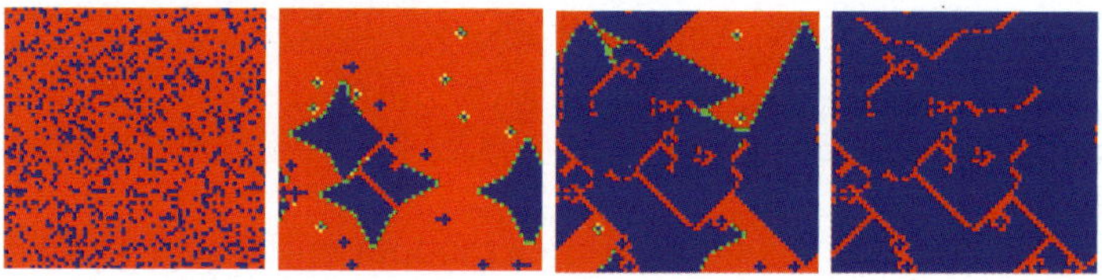

Figure C.10 The evolution of cooperation (blue) in the spatial prisoner's dilemma, starting with a random array of 80% defectors (red) using a von Neumann neighborhood. The payoffs are: $\{R; S; T; P\} = \{3; 0; 3.5; 0.5\}$.

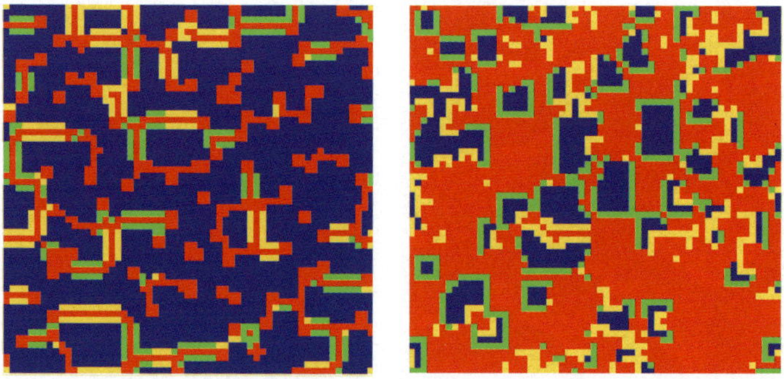

Figure C.11 Using the Nowak and May payoffs with $b = 1.6$ (left) and $b = 1.85$ (right), with an initial distribution of 10% defectors distributed randomly. Here the central cell also plays against itself using a nine-cell neighborhood. Color code is: blue = cooperator, green = new cooperator, red = defector, yellow = new defector. Periodic boundary conditions are used. Cooperation is seen to decrease with increasing values of the temptation parameter b.

Figure C.12 Employing a von Neumann neighborhood, these are three images in the time evolution of the spatial prisoner's dilemma with a value of $b = 1.4$ and an initial sole defector set in a sea of cooperators. Periodic boundary conditions are used. The color code is: blue = cooperator, green = new cooperator, red = defector, yellow = new defector.

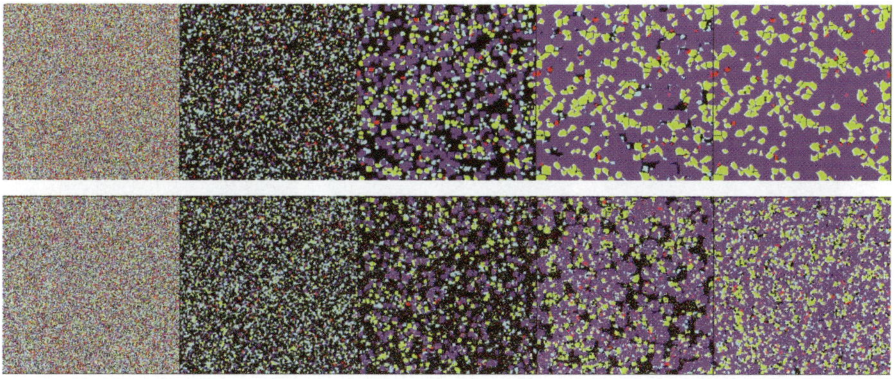

Figure C.13 The evolution of the spatial iterated prisoner's dilemma with five competing strategies with the standard payoffs from an initial random configuration. Each cell's strategy plays off five times against its eight neighbors and then adopts the strategy of its most successful neighbor. Color code: Always Cooperate = red, Tit-for-Tat = purple, Random = cyan, Pavlov = green, and Always Defect = black. The predominance of Tit-for-Tat over time is evident, with the Random strategy hardly noticeable. The second row has a certain level of noise introduced, that is, a cell will make a random move in a single round, thus increasing the dynamics of the evolution. Pavlov now holds its own against Tit-for-Tat in this instance. Images are taken at the initial state, after 2, 5, 10, and 50 iterations.

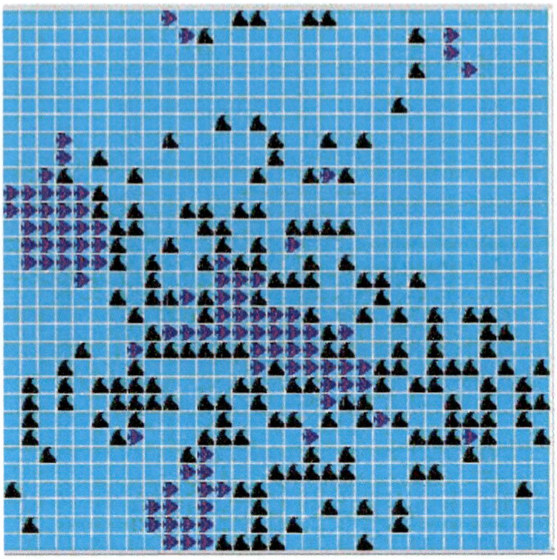

Figure C.14 The WATOR-World of fish (dark blue), sharks (black), and water (pale blue). There is an initial random mixing of 200 fish and 5 sharks. The image is taken at the 297th time step where the fish population is now 93 and the shark population has risen to 177. Breeding ages for fish and sharks is 2 and 10 generations, respectively, and the starvation time for sharks has been set at 3. The geometry of the local population mix can affect the dynamics.

Figure C.15 Stellar dendrite snow crystal forms with $\beta = 0.35$, $\gamma = 0$; $\beta = 0.4$, $\gamma = 0.001$; $\beta = 0.45$, $\gamma = 0.001$, respectively.

Figure C.16 Plate form snow crystals. The first two have the same parameter values, $\beta = 0.95$, $\gamma = 0$, and represent different time steps. Third image has $\beta = 0.95$, $\gamma = 0.0035$.

Figure C.17 The time evolution of a single avalanche (cascade of topplings) precipitated by the red cell in the first frame until a stable state is reached in the last frame. Here blue = 0, green = 1, yellow = 2, orange = 3, red = 4 and the topplings start when the site value is at least 4. The boundary was held at state 0 but is omitted from the figures.

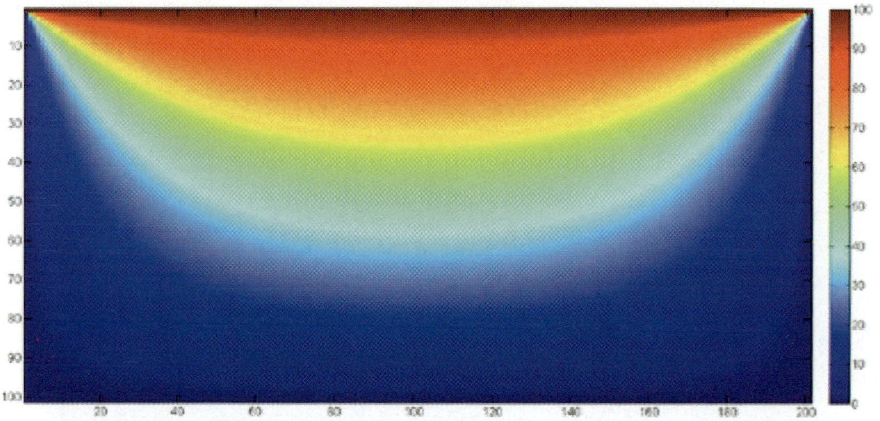

Figure C.18 Here we have color coded the steady-state temperature values of our cellular automaton simulation from red (hot) to blue (cold). See text for details.

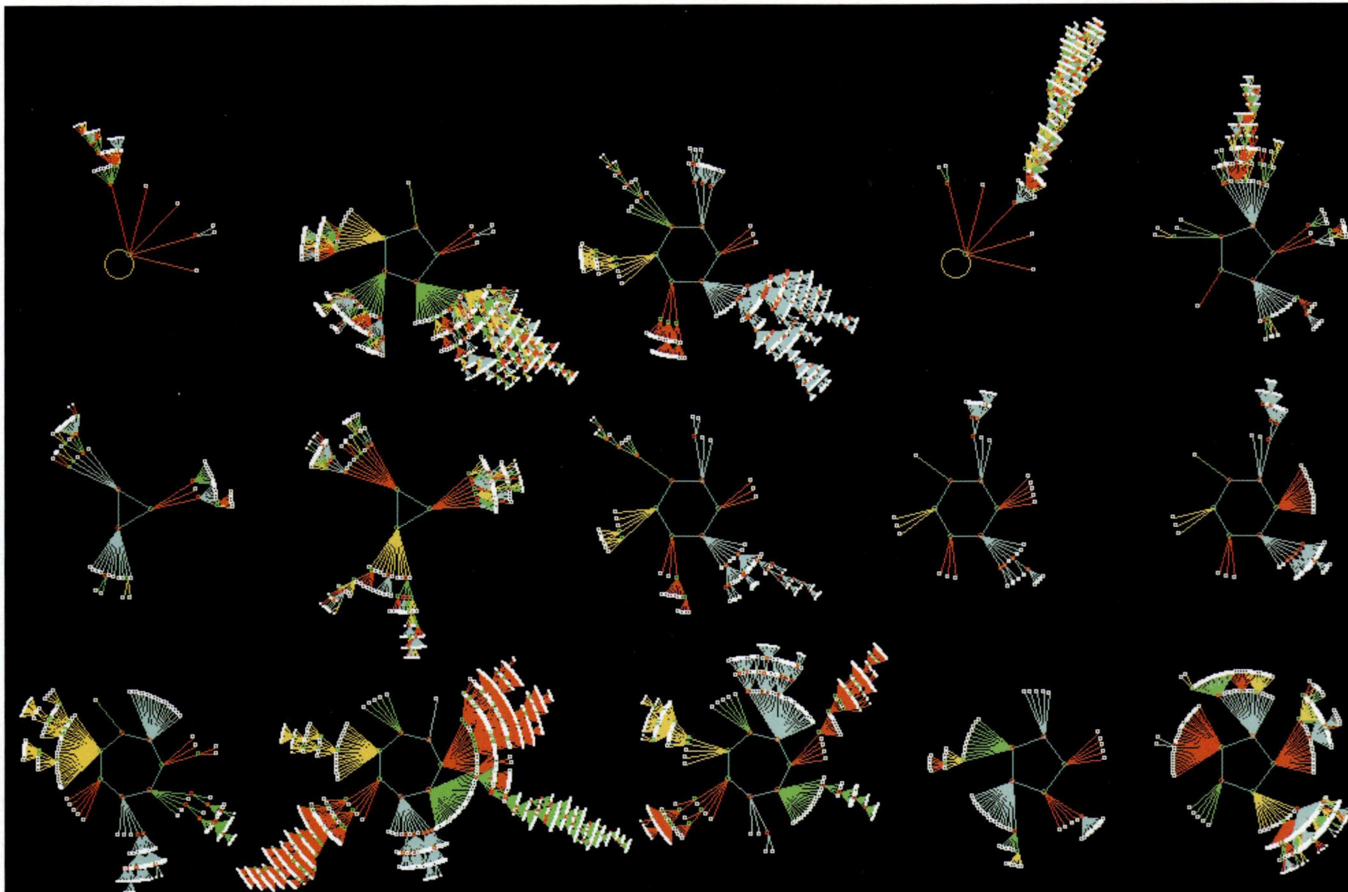

Figure C.19 These are not galaxies photographed by the Hubble Space Telescope. They are basins of attraction of an RBN with $N = 13$, $K = 3$. The state space consisting of $2^{13} = 8192$ states is organized into 15 basins having attractor periods ranging from 1 to 7.

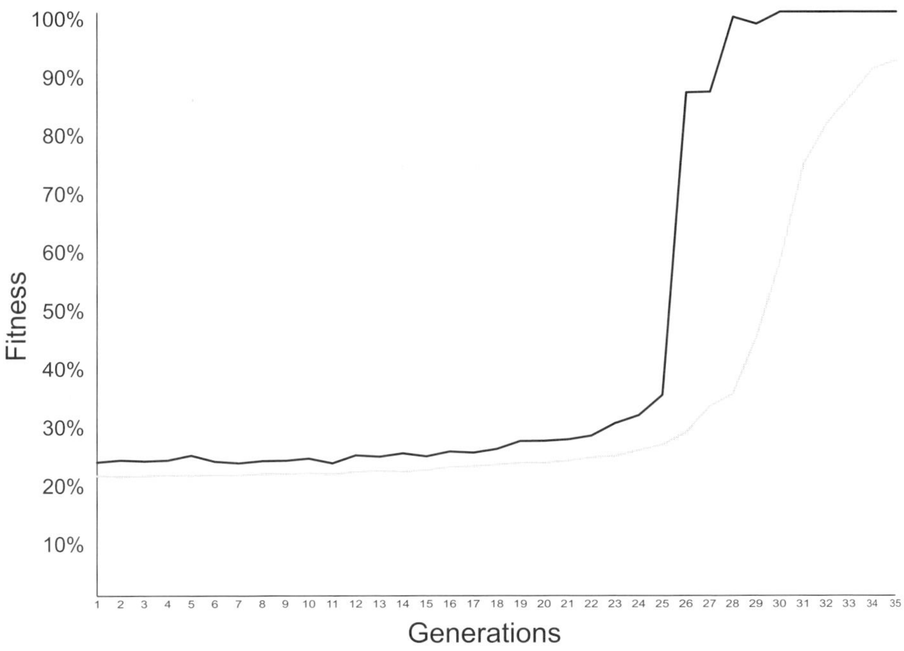

Figure 5.7 The fitness graph for the GA described in the text comparing the maximum fitness level at each generation (top black curve) to the average (bottom gray curve).

5.4.2 Predator-Prey

For many years, biologists have studied predator-prey models, in which a predator inhabits the same environment as its prey and the fluctuating populations of both are investigated subject to a variety of conditions. A set of differential equations known as the Lotke-Volterra equations has been traditionally used to describe predator-prey models based on continuous inputs. Their solution encapsulates the cyclic waxing and waning of the predator and prey populations. Since predator-prey populations are actually discrete in nature, we shall explore the popular cellular automaton model proposed by A. K. Dewdney in his *Scientific American* article [1984, 1988] known as WATOR.

Dewdney imagines a water (i.e., WATOR) world inhabited by sharks and fish where the fish are the prey, the sharks the predators, and both are allowed to reproduce. This scenario leads to five parameters which can be varied by the experimenter:

$N_f =$ initial number of fish;
$N_s =$ initial number of sharks;
$B_f =$ number of generations before a fish can breed;
$B_s =$ number of generations before a shark can breed;
$S =$ number of generations since its last fish meal before a shark starves to death.

The fish and sharks live out their lives within the framework of a two-dimensional cellular array where of course time is discrete. Let us give the water, cell state $= 0$, fish, cell state $= 1$, and sharks, cell state $= 2$. We use periodic boundary conditions so that the fish and sharks are actually swimming around on the surface of a torus (Fig. C.14, see color section). The various rules of engagement governing our aquatic inhabitants at each time step are:

- Fish swim at random to an adjacent cell in either a N,E,S,W direction provided one of these cells is unoccupied. If all adjacent cells are occupied, the fish remains where it is.
- Sharks are also able to move in a N,E,S,W direction. If some of these adjacent sites are occupied by fish, it chooses one at random, moves to that site, and eats the fish, which now departs this imaginary WATOR-World. If no fish occupy adjacent cell sites, the shark swims to one of them at random as a fish does, avoiding other sharks.
- When a stipulated number of generations have elapsed (B_f, respectively B_s), fish and sharks are allowed to breed one of their own kind. A ready-to-breed fish or shark that can swim to an adjacent cell deposits one of its offspring in the cell it is just leaving and both the offspring and adult have their biological clocks set to zero for breeding purposes. If there is no free adjacent cell for a fish, then it can neither swim nor breed. However, a shark with adjacent cells occupied by fish can still breed by choosing to eat one of the fish at random and depositing its offspring in the cell it just left. Neither fish nor sharks need to mate for the purposes of reproduction.
- Sharks will starve to death and depart WATOR-World if they have not eaten any fish after S generations. After a meal of fish, however, the life clock for the shark starts at zero again.

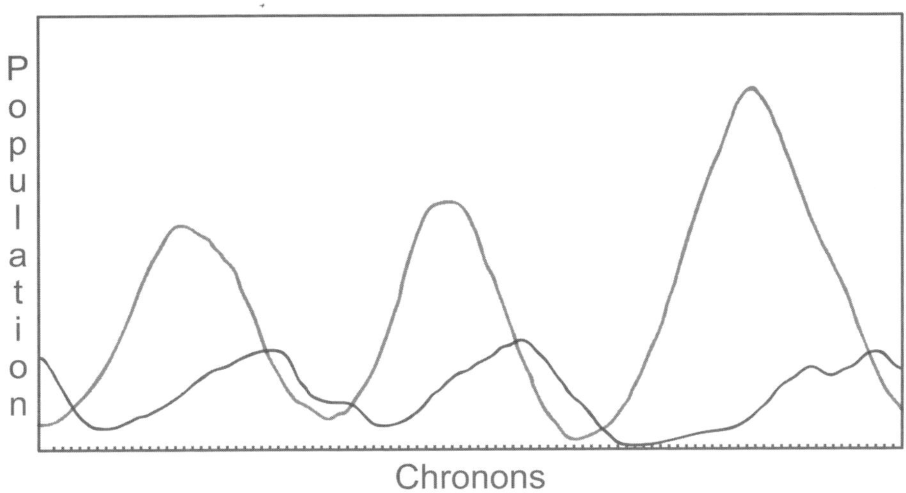

Figure 5.8 The ebb and flow of the fish (top gray curve) and shark (bottom black curve) populations from the above simulation demonstrating how the predator population follows the trend of the prey population but out of phase with it. Time steps are referred to as *chronons*, a term used by A. K. Dewdney in his *Scientific American* article.

The fish and sharks are initially distributed at random in the lattice with the vacant cells being water. In general, with a small number of sharks, the fish population increases significantly, resulting in an increase of sharks, which produces a drop in the fish population, which in turn leads to starvation and death of sharks. The remaining fish population can again increase and so on (Fig. 5.8).

However, this cyclic behavior of both fish and shark populations all depends on the spatial geometry of the two populations and the fine tuning of the five parameter values. With too many fish or too many sharks, the model soon becomes overwhelmed with one of them.

This model is a cellular automaton with probabilistic updating. Indeed, if a fish is surrounded with four empty cells in its von Neumann neighborhood, then the fate of these cells is indeterminate until the next time step when the fish chooses one at random and the fish cell and the chosen empty cell effectively exchange states.

Of course there are many variants of WATOR. Among others, if a fish or a shark cannot breed because of lack of available space, it would have to wait until the next multiple of its breed time rather than breeding at the next possible time step. This introduces a *stress factor*. One can also allow fish and sharks to swim in eight directions of the Moore neighborhood, instead of the von Neumann neighborhood employed here. Or the model can be made completely deterministic, dispensing with the random aspects altogether. In an earlier predator-prey model, Auslander *et al.* [1978] found that in one scenario the prey population exhibited chaotic behavior. It was even suggested that this could be a mechanism to prevent the predator from "tracking" the prey. On the topic of fish, an interesting probabilistic fish migration model is presented in Schönfisch and Kinder [2002].

The predator-prey model employs autonomous agents and is referred to as an *IBM* (individual-based model).

5.4.3 Bacterial Growth

Here we are interested in studying the spread of a colony of bacteria (through cell division) in the presence of a diffusable nutrient with overcrowding playing an important role. The model is due to Ermentrout and Edelstein-Keshet [1993] and uses an eight-cell Moore neighborhood. Since the word *cell* can now have two different meanings in the present context, let us stipulate that cell means a bacteria cell occupying a site (having state value 1) of a two-dimensional lattice.

The amount of nutrient in every site at the next time step is deduced from the amount at the current time step, $n(t)$, plus a weighted average of nutrient in the eight neighboring sites (see Fig. 5.35 later in the chapter) minus the amount consumed according to the formula

$$n(t+1) = (1-\delta)n(t) + \delta \cdot \text{average} - \text{eaten}$$

where

$$\text{average} = \frac{4(N+S+E+W) + (NW+NE+SW+SE)}{20},$$

(see Section 5.5.6 for a discussion of this "average"), δ is a diffusion parameter, and

$$\text{eaten} = \begin{cases} a_1 & \text{if a new cell is formed (food for growth)} \\ a_2 & \text{if a cell already exists (food for sustenance)} \\ 0 & \text{otherwise.} \end{cases}$$

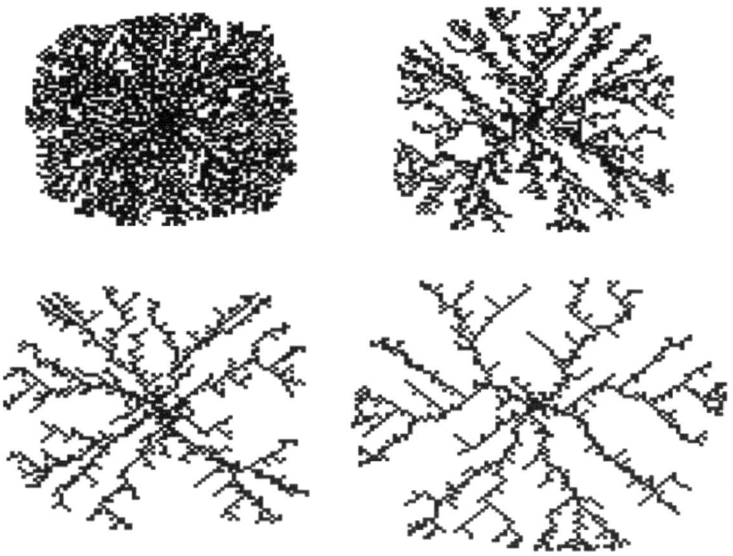

Figure 5.9 A bacteria colony and the forms it takes with increasing values of the threshold θ. Note the fractal-like appearance that develops.

Now it is necessary to take the effects of crowding into account by considering the number of surrounding sites that are occupied. This will be a function $k(t)$ for each site that takes small values for small and large numbers of occupied neighboring sites, with a suitable maximum somewhere in between. Cell growth occurs every m time steps in an unoccupied site (in state 0) with probability $\frac{1}{2}$ if for that particular site

$$k(t) \cdot n(t) > \theta$$

for some threshold value θ. The above formula means that when an unoccupied site is crowded by a lot of neighbors so that $k(t)$ is small, then a large amount of the nutrient $n(t)$ is required before cell growth can occur. A "new" cell remains as such until the next m time steps. Notice that making a_1 significantly larger than a_2 has the effect of inhibiting growth adjacent to a site where a new cell has formed as opposed to having an old cell at that site (Fig. 5.9).

This model is related to the solidification model of Packard [1986] in which higher temperature results in fewer cells experiencing the phase transition from liquid to solid. This is analogous to lower nutrient levels and lowered rates of cell growth in the bacteria model. Other features, such as inhibition and crowding, are also found in the Packard model (see Section 5.5.2).

5.4.4 Seashell Patterns

Probably most people have picked up a seashell at one time or another and marveled at its pattern of pigmentation. There are many varieties of such patterns, each with its own beauty and challenge to the scientist as to understanding its formation. It is not known why these patterns are formed as much of the life of the mollusc is spent buried in sand or mud and so they do not seem to be related to adaptation to their environment or attracting a mate. The pigmentation is secreted by individual cells along the growing edge of the mollusc and it is thought that the cells are activated and inhibited by a kind of central nervous system.

As we have already mentioned, it was Alan Turing who first worked on the problem of mathematically replicating patterns found in Nature using reaction-diffusion equations. Some years later, a computer simulation model for a pigmentation pattern on the seashell belonging to the genus *Conus* was first proposed by C. H. Waddington and Russell Cowe in 1969. Then in 1973, G. T. Herman and W. H. Liu, using their cellular linear iterative array simulator (CELIA), obtained good results of mollusc pigmentation patterns employing one-dimensional cellular automata. Wolfram (see [2002]) uses even simpler one-dimensional cellular automata to depict certain seashell patterns. A very successful simulation of seashell pigmentation patterns over a wide range of pattern types was demonstrated by Hans Meinhardt in his attractive text, *The Algorithmic Beauty of Sea Shells* [1995]. Meinhardt presented numerous sophisticated reaction-diffusion models and was able to replicate details of a large variety of shell pigmentation patterns. Young [1984] implemented reaction-diffusion models in a CA setting to reproduce patterns on animal skins. The entire topic of the use of cellular automata modeling of biological patterns can be found in the text by Deutsch and Dormann [2005].

Ermentrout, Campbell, and Oster [1986] proposed a neural activation-inhibition model which can be presented in the basic CA form

$$x_{n+1}^j = H(A_n^j - I_n^j - \theta),$$

where x_n^j represents the state (either 0 or 1) of the jth cell position in the nth row and A_n^j is the activation component, I_n^j is the inhibition component, θ is the threshold, and H is the Heaviside step function. The activation and inhibition are suitably defined in terms of weighting factors involving long-range interactions.

The cellular automata model we discuss below is due to Ingo Kusch and Mario Markus [1996] and developed out of an earlier CA model for stationary Turing patterns and spiral waves by Schepers and Markus [1992]. It is a probabalistic activation-inhibition model that uses the traditional one-dimensional array and also a circular array to permit radian growth found in some molluscs, although only the former is considered here. There are only two states for a cell $c(t)$ at time step t: 0 (resting, no pigment) and 1 (activated, pigment). Whether a cell becomes activated or deactivated at the next time step depends on the amount of chemical inhibitor present at each cell site, $I(t)$. Denote by N the number of activated cells in a neighborhood of radius r about the central cell c. Two intermediate cell states are required for both $c(t)$ and $I(t)$ in order to compute their respective values at the next time step, namely

$$x(t) \to x_1 \to x_2 \to x(t+1) \qquad \text{for } x = c, I.$$

The rules of the automaton are:

(i) **Decay of inhibitor**

If $I(t) \geq 1$, then $I_1 = I(t) - 1$, otherwise $I_1 = 0$.

(ii) **Activator production due to random fluctuations**

If $c(t) = 0$, then $c_1 = 1$ with probability p, otherwise $c_1 = c(t)$.

(iii) **Production of inhibitor due to activator**

If $c_1 = 1$, then $I_2 = I_1 + w_1$, otherwise $I_2 = I_1$.

(iv) **Activator production from diffusion of activator**

If $c_1 = 0$ and $N > \{m_0 + m_1 I_2\}$, then $c_2 = 1$, otherwise $c_2 = c_1$.

(v) **Diffusion of inhibitor**

$I(t+1) = \{\langle I_2 \rangle\}$.

(vi) **Suppression of activator due to inhibitor**

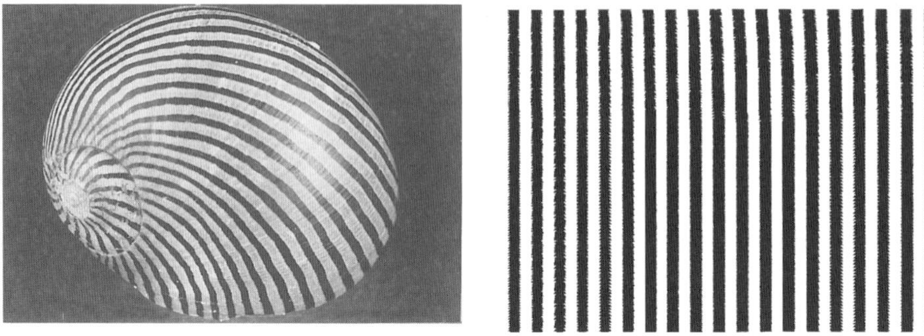

Figure 5.10 Left: *Neritina ziczac* (Philippines). Right: $r = 1$, $R = 17$, $w_1 = 1$, $w_2 = 1$, $m_0 = m_1 = 0$, $p = 2 \times 10^{-3}$.

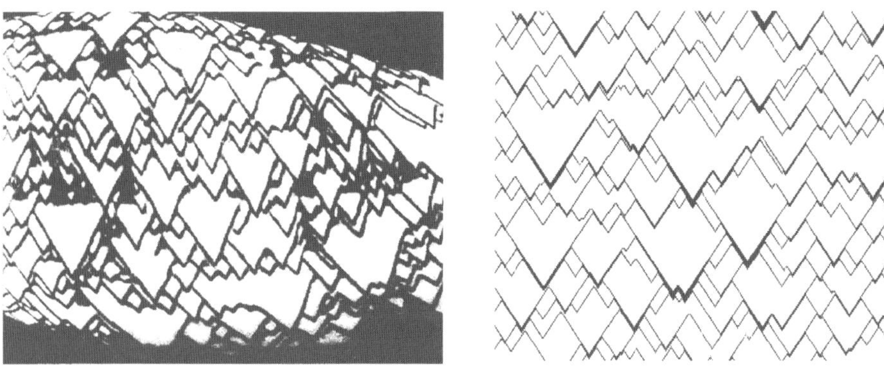

Figure 5.11 Left: *Olivia porphyria* (Panama). Right: $r = 2$, $R = 0$, $w_1 = 10$, $w_2 = 48$, $m_0 = m_1 = 0$, $p = 2 \times 10^{-3}$. The activation is propagated from cell to cell via traveling waves that annihilate upon contact with another.

$$\text{If } I(t+1) \geq w_2 \text{, then } c(t+1) = 0 \text{, otherwise } c(t+1) = c_2.$$

Here the notation $\langle \cdot \rangle$ means take the average over a neighborhood of radius R, and $\{\cdot\}$ means take the nearest integer. The authors found that at times another rule was needed which replaced rule (i):

(i*) **Decay of inhibitor**

$$\text{If } I(t) \geq 1 \text{, then } I_1 = \{(1-d)I(t)\} \text{, otherwise } I_1 = 0.$$

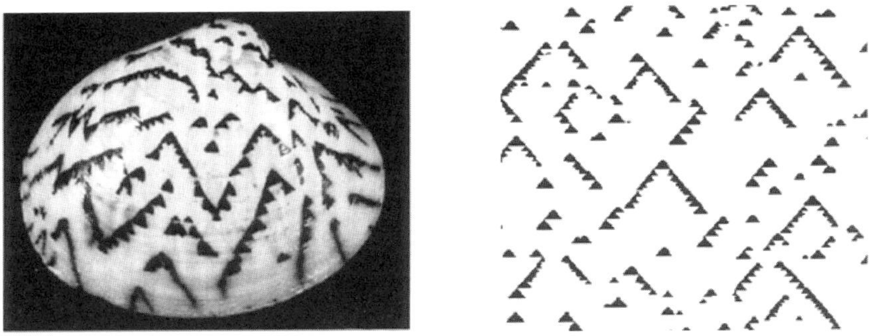

Figure 5.12 Left: *Lioconcha castrensis* (Philippines). Right: $r = 1$, $R = 16$, $w_1 = 8$, $w_2 = 6$, $m_0 = m_1 = 0$, $p = 2 \times 10^{-3}$. Considered to be of Class IV although this is not so apparent as in some other patterns.

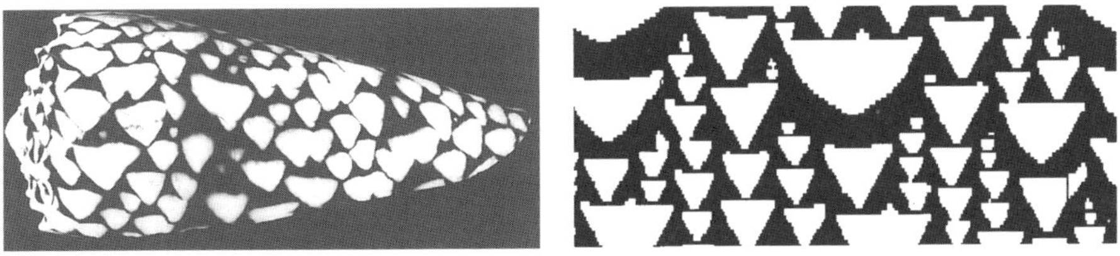

Figure 5.13 Left: *Conus marmoreus* (Philippines). Right: $r = 4$, $R = 9$, $w_1 = 2$, $w_2 = 19$, $m_0 = 0$, $m_1 = 0.2$, $p = 0$.

There are eight free parameters that the experimenter has to play with (and this activity is highly recommended): r, R, w_1, w_2, m_0, m_1, p, and d. Here R denotes the typical length for the spread of the inhibitor whose average is determined in rule (v). Also, w_1 is the amount of inhibitor produced in each activated cell per time step [rule (iii)] and w_2 is a threshold that, once it is surpassed, inactivates the cell at the next time step [rule (vi)]. The radius r is the typical length over which the activation is propagated and m_0 is the threshold for activating a cell whenever there is no inhibitor [i.e., when $I_2 = 0$ in rule (iv)], and m_1 is another activation threshold value dependent on the inhibitor value (I_2). The variable d is the fraction of the inhibitor $I(t)$ that is lost at each time step, so that rule (i*) means that the inhibitor decays exponentially, as opposed to the linear decay of the inhibitor in rule (i). Lastly, p is the probability that an inactivated cell becomes activated by random fluctuations [rule (ii)]. The automaton starts with initial conditions

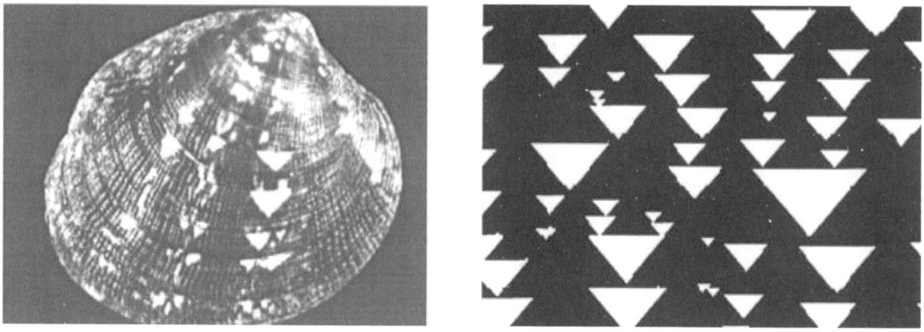

Figure 5.14 Left: *Chione grisea*. Right: $r = 1$, $R = 23$, $w_1 = 4$, $w_2 = 71$, $m_0 = m_1 = 0$, $p = 0$, $d = 0.05$ [rule (i*)].

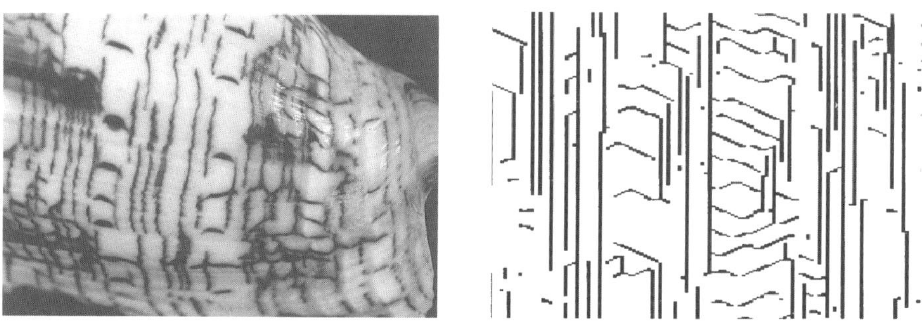

Figure 5.15 Left: *Voluta hebraea*. Right: $r = 1$, $R = 2$, $w_1 = 5$, $w_2 = 10$, $m_0 = 0$, $m_1 = 0.3$, $p = 2 \times 10^{-3}$. Stationary lines (where the inhibitor is high) and traveling waves (where the inhibitor is low) are the two competing forms. Considered to be of Class IV.

$I(0) = 0$ for each cell site with the activated and nonactivated cells distributed randomly.

Below are some samples of various species of mollusc and their cellular automaton counterparts together with the parameters used in each case.

One very noteworthy feature of this model is that the pigmentation patterns generated by it include the Wolfram classes II, III, and IV. In Fig. 5.10 we see an example of Class II behavior, and in Figs. 5.11, 5.13, and 5.14, Class III dynamics is evident. In Figs. 5.15 and 5.16 we see evidence of Class IV phenomena, and in their paper Kusch and Markus have even devised a quantitative measurement which is used to discriminate between the four classes to support their assertions. Accordingly, Fig. 5.12 is also given Class IV status, although this is not

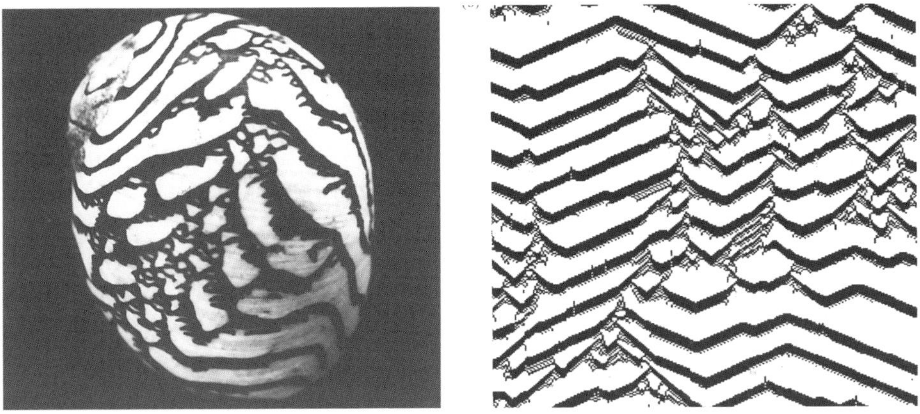

Figure 5.16 Left: *Neritina pupa* (Antilles). Right: $r = 2$, $R = 0$, $w_1 = 6$, $w_2 = 35$, $m_0 = 0$, $m_1 = 0.05$, $p = 2 \times 10^{-3}$, $d = 0.1$ [rule (i*)]. Considered to be of Class IV.

so apparent visually. The authors claim these examples "would be the first experimental evidence for class IV processes in nature."

It is worthwhile to compare the CA approach of Kusch and Markus with the reaction-diffusion models of Meinhardt. As it turns out, some shell patterns are modeled equally well in both approaches, some are more effectively simulated using the CA approach (those involving traveling waves and Class IV patterns), while it is conceded that the Meinhardt-Klinger partial differential equations have managed to reproduce some patterns that the CA could not. At least not yet.

5.5 PHYSICAL MODELS

Cellular automata have been used to model many of the dynamic features of the physical world. In this section we will discuss just a few of the more notable applications.

5.5.1 Diffusion

The dispersion of heat or of one chemical substance into another due to the random movement of molecules has classically been modeled by the diffusion equation, which in one dimension is given by

$$\frac{\partial u}{\partial t} = c\frac{\partial^2 u}{\partial x^2},$$

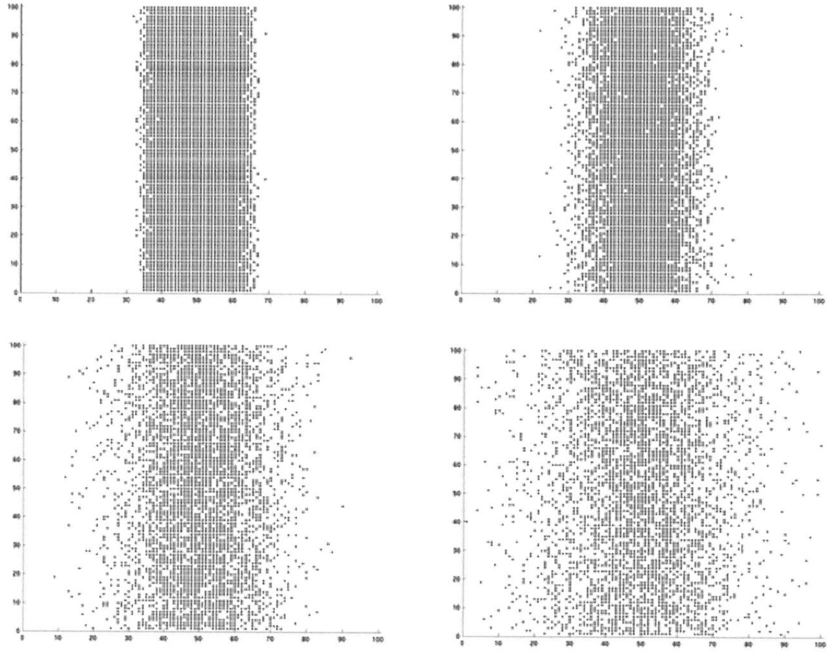

Figure 5.17 The diffusion of a solid substance after 100, 1000, 5000, and 10,000 steps.

where $u = u(x, t)$ is the amount of heat or concentration of the dispersing substance at distance x and time t and c is just a constant that depends on the medium. One shortcoming of the solution is that at any given distance from the source, there is to be expected some quantity of the substance at any given time. So, for example, if you throw a lump of sugar into a warm bathing pool, the diffusion equation solution of this event always predicts a small quantity of sugar far from the site of the initial lump even a short time after it enters the water.

By contrast we can produce a cellular automaton model that is much closer to what one expects as in Figs. 5.17 and 5.18.

The automaton employed here is quite simple and tends to mimic the actual diffusion of a substance like sugar in water in which particles of sugar swap places with parcels of water. In our example, we are taking 100 rows with the solid substance such as sugar, represented by the vertical strip of cells in blue which can diffuse into the surrounding medium of water represented by white cells. Taking one row at a time,

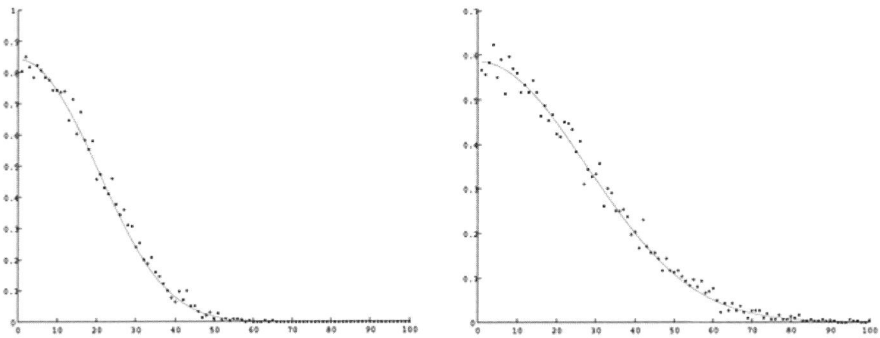

Figure 5.18 The density values (vertical axis) for $u(x,t)$ for $t = 100$ (10,000 iterations) and $t = 300$ (30,000 iterations), respectively, plotted with the theoretical values of u given by the solution to the diffusion equation. The diffusion of the material into the surrounding medium is evident. Only the right-half of the bar is considered.

we choose a cell at random and swap its contents (i.e., state) with the adjacent cell to its left. We then proceed to do this for each of the 100 rows and count this as the first time step. Then we start again with the first row, choosing a cell at random and swapping its contents with the adjacent cell to the right, again carrying out this procedure for each of the rows. This is the second time step. We continue in this fashion so that at each odd time step (1,3,5,...) we swap the contents of a random cell in each of the rows with the adjacent cell to the left, and at even time steps (2,4,6,...) we swap with the cell to the right. If two cells have the same contents, then their swapping will result in no change to either.

Quantitative data can even be extracted from the cellular automaton model by choosing a particular vertical column, that is, a distance x from the source, and adding up the number of occupied cells n. Then the average, $n/100$ (since there are 100 cells vertically), is a density value for $u(x,t)$ for the given time t representing, say, the amount of sugar present at the given distance and time.

5.5.2 Snow Crystals

Snow crystals form when water vapor condenses on some seed, usually a speck of dust. When water freezes, molecules of water ice arrange themselves in a hexagonal lattice, the result being that ice crystals take the form of a hexagonal tablet or prism. At certain temperatures the sides grow much more rapidly than the top and bottom surfaces and the crystal spreads out into a planar figure. Since the snow crystal experiences the same ambient conditions across its entire face, it grows with a sixfold symmetry. However, the crystal growth is strongly influenced by the temperature and as the snow crystal is blown about it experiences different temperatures which lead to different morphological features. Thus the physics of how water molecules are added to the growing crystal structure can be very complex. By the way, snowflakes are actually assemblages of snow crystals that are loosely bound together.

The first scientific musings on snow crystals, and indeed on morphogenesis in general, goes back to Johannes Kepler's short monograph, *The Six-Cornered Snowflake*. Kepler was interested to know why snow crystals always exhibited sixfold symmetry and attributed it to a force he termed *facultas formatrix*, or "formative faculty," found in all matter. It would be another 300 years before an intensive investigation of snow crystal formation would be undertaken by the Japanese scientist Ukichiro Nakaya, who was able to grow snow crystals in his laboratory under varying atmospheric conditions. Nakaya photographed and classified the different snow crystal forms and produced a famous diagram plotting the different forms against values of temperature and humidity.

One of the pioneers of snowflake photography was Wilson Bentley (1865–1931), a Vermont farmer who turned to snow crystal photography taking more than 5000 images in his lifetime (see Fig. 5.19). His book, *Snow Crystals*, published in 1931, has been rereleased by Dover [1962] and is still in print.

Norman Packard [1986] has shown that a simple automaton model can reproduce many of the snow crystal patterns quite effectively. When a water molecule is added to the crystal structure, a small amount of heat is released at the site which acts as an inhibiting factor for more water molecules to attach at the same site. So the Packard automaton model states that:

- A cell site becomes black (representing the presence of solid material) if it has exactly one black neighbor.
- A cell remains white (representing the absence of solid material) if it has more than one black neighbor.

Figure 5.19 Some of the exquisite forms of actual snow crystals. Photographs taken by Wilson Bentley.

Starting with a single black cell (the seed) on a hexagonal lattice, we obtain a sequence of cellular automata snow crystals (Fig. 5.20).

Very clear similarities can be seen between the actual snow crystals above and the cellular automata ones. One can also note an evolution that alternates from a branched structure to the tablet form and back again.

A more sophisticated model capable of capturing snow crystal forms has been presented by Clifford Reiter [2004]. This model also employs a hexagonal grid and states of cells are allowed to take on real number values as opposed to simply integer values. Such models with states allowed to take real values are referred to as *continuous automata*. In this instance the values are a measure of the amount of water at a particular cell site, and if that value is ≥ 1 then the water is taken to be ice. The cellular array is divided into two distinct classes that are updated differently:

- Receptive (**R**): cells that are ice (value ≥ 1 or have an immediate neighbor that is ice.
- Unreceptive (**UR**): all other cells.

Figure 5.20 The evolution of the Packard model of snow crystal formation starting with a seed of a single black cell on a hexagonal lattice. A close correlation with actual snow crystals is observed.

The **R** sites are updated at the next time step by just adding a constant parameter γ, the idea here being that water may be supplied from outside the plane of growth and stored.

The **UR** cell sites are not yet ice and so are governed by the diffusion of water vapor. If $u = u(P, t)$ represents the amount of water vapor at position P and time t, then the diffusion equation describing the vapor flow is

$$\frac{\partial u}{\partial t} = a\Delta u,$$

where $\Delta u = \partial^2 u/\partial x^2 + \partial^2 u/\partial y^2$ is the Laplace operator applied to u and a is constant. On a hexagonal array, this can be expressed in a discrete form as

$$u(P, t+1) \approx \frac{u(P,t)}{2} + \frac{\sum_{N \in \mathrm{nn}(P)} u(N,t)}{12},$$

where $\mathrm{nn}(P)$ represents the set of nearest neighbors of P, and we have taken $a = \frac{1}{8}$ as the canonical case.

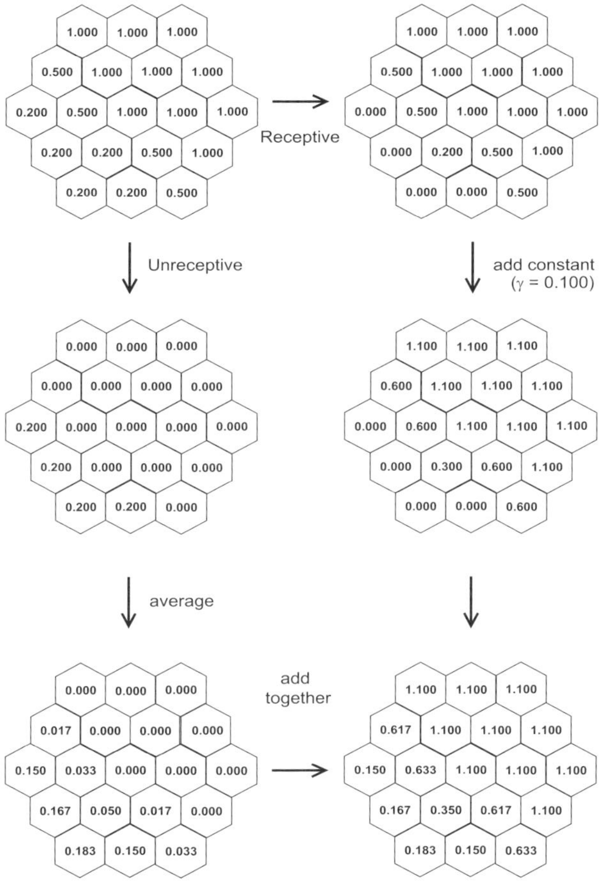

Figure 5.21 Updating the hexagonal array of receptive and unreceptive cell sites as described in the text.

The updating is a two-stage process. First, the array is split into receptive and unreceptive groupings. The receptive group consists of all the **R** cell sites but also has 0 as a *place holder* at any **UR** cell sites. The unreceptive group consists of all the **UR** cell sites with 0 as a *place value* for the purposes of averaging at any of the **R** cell sites. See Fig. 5.21.

As mentioned above, all the **UR** cell sites in the receptive group have the constant γ added to their value with 0 placeholding the **UR** sites. All the sites in the unreceptive group are updated according to the averaging given by the preceding equation, even the sites having value 0. Finally, the corresponding cell values of the receptive and unreceptive groups

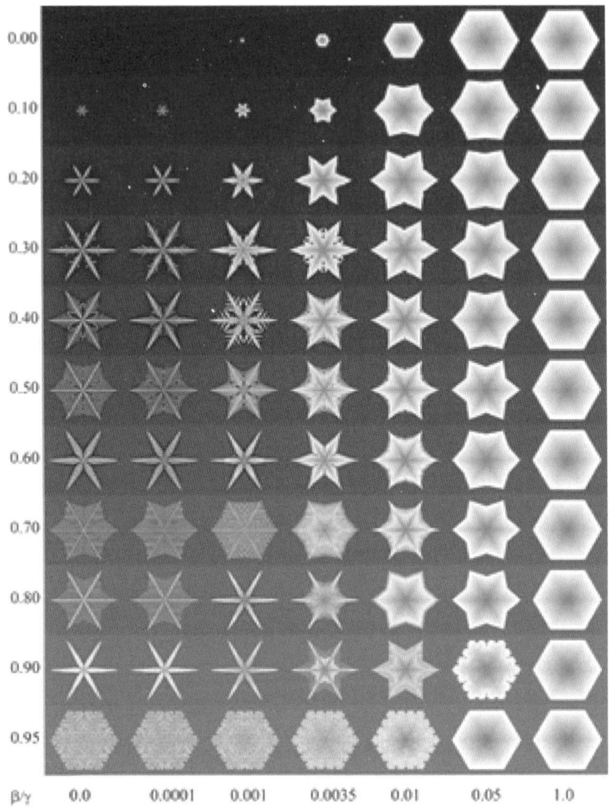

Figure 5.22 The morphological changes of the snow crystals in the Reiter model as a function of the background parameter β and the additive constant γ.

are added together and this gives the state of the array at the next time step.

There is one other parameter that we are free to set and that is the background level given by the constant β. By a suitable choice of the two parameters γ and β, the model is able to generate a large variety of snow crystal forms. Of course we could also vary the parameter α, but Reiter, has found that in general the qualitative behavior of snow crystal growth is not affected by the choice of α. In Figs. C.15 and C.16 (see color section for both) we start from a single seed of value 1 with all other cells having background value β.

The morphological changes with varying β and γ can be seen in Fig. 5.22.

Beautiful snow crystal images can be found on Kenneth Libbrecht's website `www.its.caltech.edu/~atomic/snowcrystals/`. Once you have seen these, you will be sorely tempted to obtain the book by Libbrecht and Patricia Rasmussen [2003].

5.5.3 Sandpile Model

Many of us have had the childhood pleasure of building sandpiles at the beach. But in 1988, Bak, Tang, and Wiesenfeld made them an object of scientific investigation, and piles of sand (both real and ideal) are now being played with by adults as well. In what follows we will consider only a single idealized case in a two-dimensional CA setting, although a considerable literature has developed arising from the original seminal study (see Sornette [2004, Section 15.2]).

The lattice will be square, unlike a more conventional sandpile which has a circular base, but this is not significant. Adding a grain of sand to a cell site increases the state value of the site by 1. Boundary sites are maintained at the value 0, which represents sand falling off the edge and out of the model. Sand is then added, one grain at a time, to any cell site at random. As sand begins to accumulate at cells in the lattice, we wish to emulate the toppling of sand (release of tension) when the height at a particular site reaches a threshold value and becomes unstable:

- Whenever a site has reached a height of at least four grains of sand (i.e., state value ≥ 4), the site value is then reduced by 4 and the four sand grains are redistributed to each of the four von Neumann nearest neighbors, which now have their state value increased by 1.

The above toppling mechanism may precipitate further toppling in some of the neighboring sites, which in turn can produce a cascade of topplings (called an *avalanche*) until all the lattice sites have become stable (Fig. C.17, see color section). It is only then that another grain of sand is added randomly to a cell site on the lattice. Starting from a stable initial condition, each cycle adds a grain of sand randomly to the lattice and allows the ensuing cascades to unfold and reach a stable configuration before another cycle begins.

There are several variations of the sandpile model, and the one considered here is the classical abelian sandpile. The term abelian means *commutative*, as in the commutative property of addition: $a+b = b+a$. In our sandpile context, it means that if a grain is added to a stable

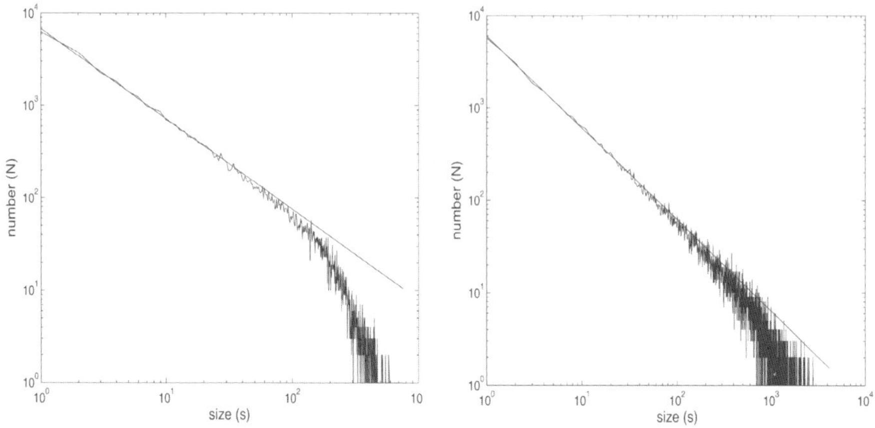

Figure 5.23 Log-log plots of the number of avalanches N vs. size s. The slope of the straight line is -1. The left-hand graph used a grid of 20×20 and the right-hand grid was 40×40. Notice that grid size does matter.

configuration at site i and then a grain is added at site j, this results in the same stable configuration as adding the grains to the sandpile sites in the reverse order.

After some considerable number of cycles, the sandpile reaches a so-called critical state in which the addition of a single grain of sand results in avalanches of all sizes and extending over all time durations, constrained of course by the size of the system. The statistical dynamics of this particular model have been extensively studied and it has become the prototype for the notion of *self-organized criticality* (*SOC*) by Bak *et al.*, in which the critical state is reached without tuning any external parameter like temperature as in the Ising spin model (Section 5.5.5).

In a computer experiment, lattices of 20×20 and 40×40 were employed and allowed to reach a critical state. Then the number of avalanches, N, was plotted on a log-log graph against their *size* s, which is determined by the number of topplings comprising each avalanche (Fig. 5.23). A power law relationship, $N \propto s^{\tau}$, is indicated, with $\tau = -1$ being the slope of the straight line. The rapid fall-off of the largest avalanches is seen to be an artifact of the finite size of the model and, of course, larger arrays must be considered if we are to believe that this is a true power law operating at all scales.

It should be noted that real sandpiles are rather less well behaved than our idealized one and evidence for SOC appears to be dependent upon taking a small size for the horizontal circular plate supporting the sandpile (Held *et al.* [1990]). Moreover, SOC has been claimed to have been found in other cellular automata, such as in the Game of Life and so-called forest fire models, but even in these idealized cases finding SOC at all scales is rather problematic, and in the latter case, as Grassberger [2002] points out, power laws are only transient phenomena. The author cautions, "Log-log plots have a notorious tendency to suggest linear curves, even if there are no real power laws." Nevertheless, the notion of self-organized criticality, or *self-organized complexity*, as some researchers prefer to call it (Newman and Turcotte [2002]), will continue to provide a rich field for scientific study.

5.5.4 Lattice Gases

Our first model for the simulation of fluid flow is due to Hardy, de Passis, and Pomeau [1973] and accordingly is known as the HPP model. It is a two-dimensional model of fluid particle interactions, where time is discrete and particles are all identical and travel at unit speed. The lattice is rectangular and particles are deemed to be at the nodes of the lattice and can travel only in the four directions of the lattice itself. One's first impression is that this is not really a cellular automaton, but as we shall see it turns out that it is equivalent to the partitioning cellular automata of Margolus. Lattice gas models also appear in biology models (cf. Ermentrout and Edelstein-Keshet [1993]) since the particles can also represent cells or organisms.

At each time step, each particle travels from one node to the next nearest node, as indicated by its velocity vector (which is just an arrow indicating the direction of flight along one of the lattice lines). An exclusion principle forbids any two identical particles to occupy the same node at the same time, but up to four particles can collide at the same node. The rules for interaction between particles at the same node are the following:

- Particles arriving at a node from opposite directions (head-on collisions) give rise to two particles traveling at right angles to the original direction of flight (Fig. 5.24, left).

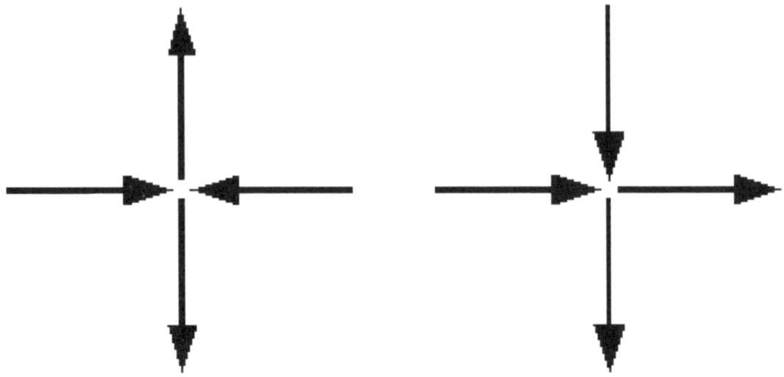

Figure 5.24 Particle interactions in the HPP model. Head-on collisions result in particles exiting at right angles (left) and any other type of collision results in the particles continuing on their original course.

- In all other particle collisions (say at right angles or between more than two particles) the particles simply continue on the path indicated by their velocity vector (Fig. 5.24, right).

So that the above collision rules make sense when thinking about collisions, say, between billiard balls, it is important to note that individual particles cannot be distinguished after a collision takes place. There are actually 16 possible interactions at a node which we may describe in the following table, where we have indicated the directions of flight by N (north), W (west), S (south), E (east) with a 1 denoting the presence of a particle and a 0 the absence of one:

In					Out						In					Out			
N	W	S	E		N	W	S	E			N	W	S	E		N	W	S	E
0	0	0	0		0	0	0	0			1	0	0	0		1	0	0	0
0	0	0	1		0	0	0	1			1	0	0	1		1	0	0	1
0	0	1	0		0	0	1	0			1	0	1	0	*	0	1	0	1
0	0	1	1		0	0	1	1			1	0	1	1		1	0	1	1
0	1	0	0		0	1	0	0			1	1	0	0		1	1	0	0
0	1	0	1	*	1	0	1	0			1	1	0	1		1	1	0	1
0	1	1	0		0	1	1	0			1	1	1	0		1	1	1	0
0	1	1	1		0	1	1	1			1	1	1	1		1	1	1	1

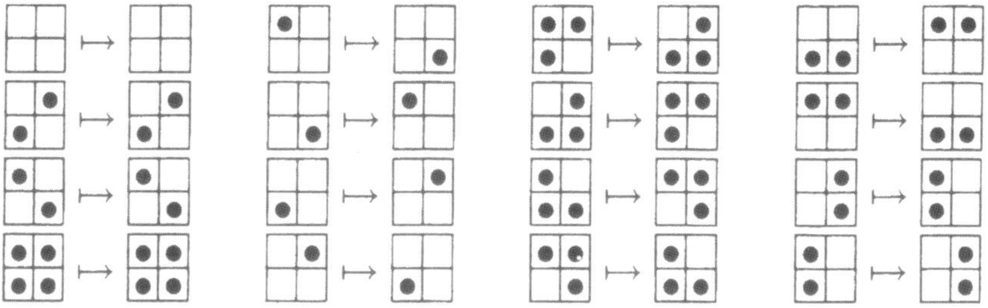

Figure 5.25 SWAP-ON-DIAG: In every 2×2 of the current partition, swap the contents of each cell with that of the cell diagonally opposite to it in the block.

Only the two head-on collision events indicated by (*) produce any change in direction for the incoming particles. Observe that the number of incoming particles (which are identical in mass and velocity) always equals the number of the outgoing particles so that energy and momentum (mass $\times$ velocity vector) are conserved by collisions. The HPP model captures many of the macroscopic properties of fluid dynamics. However, one peculiarity of this model is that the horizontal component of momentum is conserved along all horizontal lines of the lattice, and the vertical component of momentum is conserved along all vertical lines. As this does not actually happen in the real physical sense, it leads to discrepancies.

However, let us see how the HPP model follows from the Margolus neighborhood notion. Following Toffoli and Margolus [1987], let us adopt just the simple rule that the authors call SWAP-ON-DIAG (see Fig. 5.25).

By considering, say, a particle in the bottom left-hand corner of a dark 2×2 block, it will, according to the SWAP-ON-DIAG rule, go diagonally to the top right-hand corner of the block. Next it will be considered in a light block, again at the bottom left-hand corner, and the rule will move it further diagonally to the top right, as in Fig. 5.26 (left). Particles at other positions will also move diagonally.

At this stage the SWAP-ON-DIAG rule has produced particle motion, but not any collisions and particles heading directly toward one another will just by-pass one another, traveling merrily on their way, as in Fig. 5.26 (right).

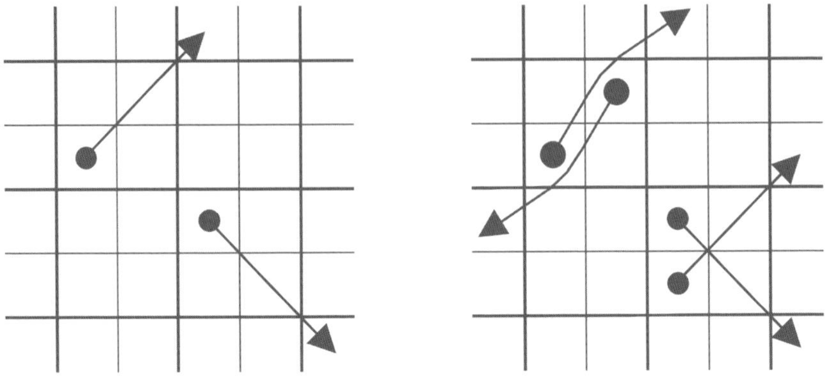

Figure 5.26 Particle movement following the SWAP-ON-DIAG rule. Movement is always diagonally across the lattice (left). Particles that encounter one another, either head-on or side-on move as if "through" one another (right).

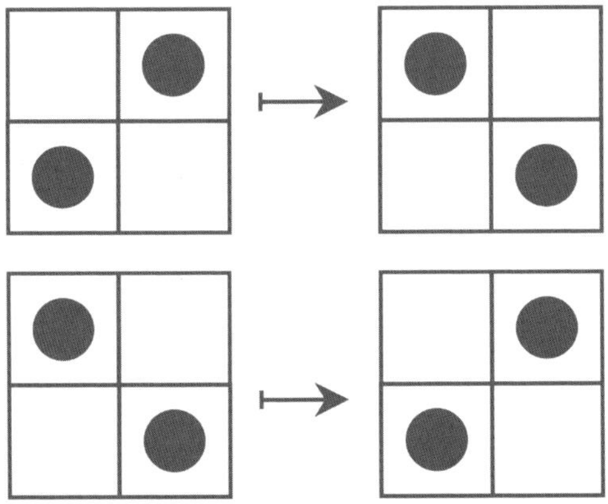

Figure 5.27 The two amended rules in SWAP-ON-DIAG to produce the head-on collision behavior of the HPP model.

In order to account for the collision rule of the HPP model we need only make one slight emendation to the SWAP-ON-DIAG rule. Namely, where there are two particles at opposite ends of a diagonal in a block (and no other particles in the block!), we ROTATE the two particles 90 degrees to the other diagonal. Thus we simply need to amend the second and third rules of SWAP-ON-DIAG as in Fig. 5.27 and maintain all the others as they are.

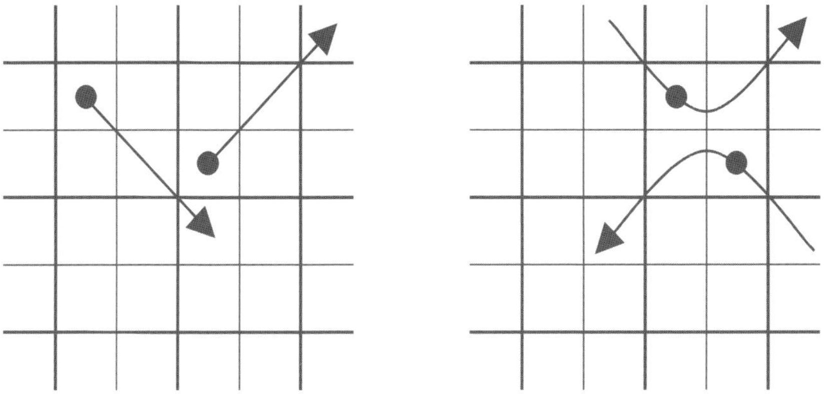

Figure 5.28 Particle movement with the modified SWAP-ON-DIAG rule showing free movement in space (left) and a head-on collision (right).

By way of illustration, in Fig. 5.28 we have the movement of two particles moving freely through space (left) and colliding head-on (right).

Thus all the dynamics of the HPP lattice gas model have been reproduced using the Margolus neighborhood and the preceding modification of the SWAP-ON-DIAG rule.

A model employing the same general principles but which provides a much better fit with real fluids is due to Frisch, Hasslacher, and Pomeau [1986] and is known as the FHP model. It uses a hexagonal lattice (again in two dimensions) so that particles can travel along six different directions. A sample of particle interactions is given in Fig. 5.29. Both mass and momentum are again conserved at each vertex.

The virtue of the FHP model is that in the macroscopic case taking a sufficiently large number of particles, the classical two-dimensional Navier-Stokes equations are satisfied. The Navier-Stokes equations are the fundamental partial differential equations governing the fluid flow of liquids and gases. For some additional technical discussion regarding the behavior of the viscosity of the model, see the references of Frisch *et al.* [1987, 1990]. Various other lattice gas models have been devised, including one for three-dimensional hydrodynamics (Henon [1987]).

5.5.5 Ising Spin

An idealized system that originally arose out of a study of ferromagnetic materials, has become one of the canonical models of modern

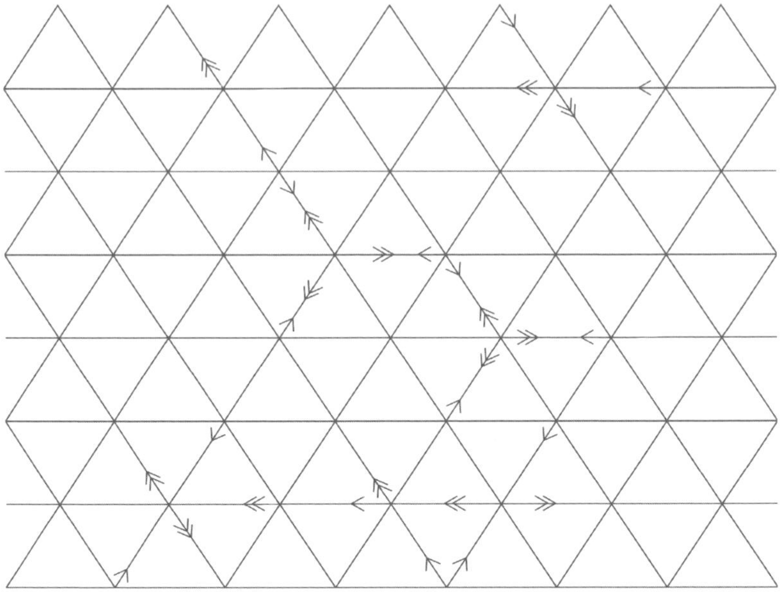

Figure 5.29 Various particle interactions in the FHP model. Particles at time t are given by a single arrow, and those at time $t+1$ are given by double arrows. Head-on two-body collisions have two possible outcomes with equal *a priori* weights (bottom left and bottom right). Particle trajectories for other interactions not depicted are unaffected by collisions.

physics. An array of spins was proposed by Wilhelm Lenz in 1920 and investigated mathematically by his student Ernst Ising. The resulting model considers a lattice of sites at each of which is placed a *spin* which takes on either the value $+1$ or -1. Equivalently, one can think of the $+1$ spin as an arrow $[\uparrow]$ representing *spin up* and the -1 spin as an arrow $[\downarrow]$ representing *spin down*. The lattice we consider in the sequel consists of squares, but triangular or hexagonal lattices have also been employed in two dimensions, although the lattice can be in any dimension. The Ising system also serves to model a large class of thermodynamic systems that experience phase transitions.

The spins lie at the vertices of the lattice which results in an interaction energy (or *bond energy*) between each site's spin and those of its von Neumann nearest-neighbor sites. The energy contributed by these interactions is $+1$ whenever the sites have *opposite* spins and -1 whenever the sites have the *same* spin, in the sense that opposite poles

$$
\begin{array}{ccccccc}
[\downarrow] & -1 & [\downarrow] & +1 & [\uparrow] & +1 & [\downarrow] \\
+1 & & +1 & & +1 & & +1 \\
[\uparrow] & -1 & [\uparrow] & +1 & [\downarrow] & +1 & [\uparrow] \\
-1 & & -1 & & -1 & & -1 \\
[\uparrow] & -1 & [\uparrow] & +1 & [\downarrow] & +1 & [\uparrow] \\
-1 & & +1 & & -1 & & +1 \\
[\uparrow] & +1 & [\downarrow] & -1 & [\downarrow] & -1 & [\downarrow]
\end{array}
$$

of a magnet attract each other and like poles repel. This is illustrated in the example below.

Here the spins up $[\uparrow]$ and spins down $[\downarrow]$ each have four bond energies with their nearest neighbors in the NESW directions, which are $+1$ when the spins are opposite and -1 when the spins are the same.

In order to compute the total energy of the system, one adds up all the nearest-neighbor energies for the lattice. If we denote by s_i ($=\pm 1$) the value of the spin at site i, then the bond energy between the nearest-neighbor pair s_i and s_j is given by the negative of their product: $-s_i\, s_j$. Taking the sum over all of these nearest-neighbor pair interactions gives the total magnetic energy of the system:

$$\mathcal{E} = \frac{1}{2}\sum_{i,j} -s_i\, s_j = -\frac{1}{2}\sum_{i,j} s_i\, s_j.$$

Periodic boundary conditions are taken to achieve the closest approximation to an infinite lattice. In order to have an energy value that does not depend on the size of the system, we will normalize it by dividing by the total number of spins N and denote $E = \mathcal{E}/N$. In the extreme cases when *all* the spins are either up $(+1)$ or down (-1), we find that $E = -2$ since there are four bond energy sums for each of the N spin sites with each having a value of -1 and the factor of $\frac{1}{2}$ compensates for the fact that we have counted each bond energy twice, giving $E = -2N/N = -2$. In the other extreme case when all the spins alternate consecutively between up $(+1)$ and down (-1) over the entire lattice, there are $4N$ bond energy sums each having the value $+1$ so that $E = 2$. Thus the normalized energy has the bounds $-2 \leq E \leq 2$.

The magnetization of the system is given by the sum of all the values s_i and normalized again by dividing by the number of spins N, giving

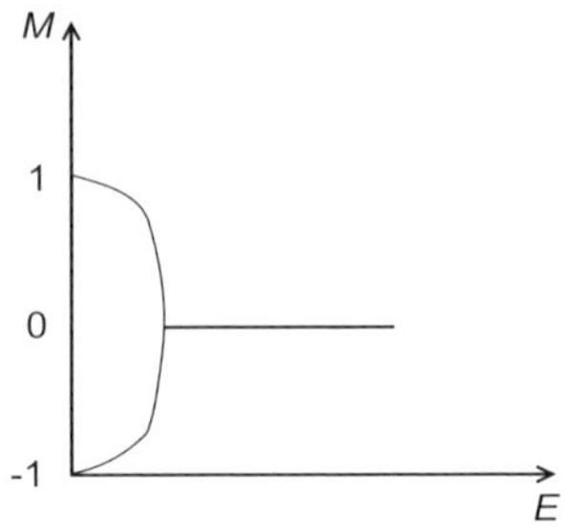

Figure 5.30

$$M = \frac{1}{N} \sum_i s_i. \tag{5.1}$$

This means that if all the spins were up $(+1)$ then $M = +1$ and if all the spins are down then $M = -1$, so that the magnetization lies in the range $-1 \leq M \leq 1$. The formulation for magnetization in Equation 5.1 also represents M as the fraction of the up spins minus the fraction of the down spins and hence $M = 0$ when exactly half of the spins are up and half are down.

Let us just take note of what we have created here. For an array that is 10×10 so that $N = 100$, the total number of different lattice configurations equals $2^{100} \approx 10^{30}$, a rather large number indeed. The set of all these configurations is called the *phase space*. This is the universe we get to roam around in to explore the dynamics of magnetization. Taking the extreme case again when all the spins are either all up $(+1)$ or all down (-1), we found that $E = -2$ and either $M = 1$ or $M = -1$, respectively. Even when a large proportion of the spins are either all up or all down, the energy E will be low and close to its minimum. Furthermore, from the model we can glean the fact that whenever exactly half the spins are either up and half are down, $M = 0$ as mentioned above. Moreover, there are many different possible configurations with half the spins up and half the spins down (essentially where there is a random ordering of the spins) and each one has a possibly different total energy. We can conclude that for a large number of values of E we will have $M = 0$. This includes the case when the spins alternate consecutively up and down and the energy has the maximum value $E = 2$. It turns out that the limiting situtation as the array becomes infinite in extent looks like Fig. 5.30 when we graph the magnetization *versus* energy for different configurations.

In a physical context, it is observed that the magnetization of a magnetic material is influenced by its temperature. That is, at high temperature, there is essentially no magnetization, and as the temperature is lowered below some critical (*Curie*) temperature, T_c, regions of magnetization are spontaneously generated. The high-temperature scenario corresponds to a high internal energy and a random ordering of the spins (essentially half the spins up, half down), so that M is effectively 0, whereas in the lower temperature case, spins tend to orient in a particular direction over regions wherein M is either positive or negative. This spontaneous generation of magnetism is a phase transition and is represented by a kink in the behavior of M at T_c (actually, a discontinuity in the rate of change of M), as can be seen in the graph. Ising had shown in his 1924 thesis that no phase transition arises in the one-dimensional model. Once a neighbor of Albert Einstein, Ising died in 1998 having taught physics for nearly three decades at Bradley University in Illinois.

In 1944, it was demonstrated by Norwegian chemist, Lars Onsager, that in the two-dimensional case there is a complete mathematical description of the phase transition in the Ising spin model. Onsager went on to win the Nobel Prize in Chemistry in 1968 for his work on a mathematical description of irreversible processes.

In order to explore the dynamical behavior of the Ising model since there is no time parameter in the model itself, the system is made to evolve according to some algorithm that will "flip" the spins in some fashion. We discuss an elementary formulation that preserves the total energy of the system that is based on the so-called Q2R rule of Vishniac [1984] and developed in Toffoli and Margolus [1987]. The algorithm is: *we flip all and only spins that have two nearest neighbors with spins* $+1$ *and two with spins* -1, so that the sum of the four bond energies associated with each such spin remains unchanged. However, one must be a bit careful how the flipping of spins is carried out. In the figure below, the two central (down) spins can both flip according to our rule,

$$\begin{array}{cccccc} & \cdots & [\uparrow] & [\uparrow] & \cdots & \\ \cdots & [\uparrow] & [\downarrow] & [\downarrow] & [\downarrow] & \cdots \\ & \cdots & [\downarrow] & [\uparrow] & \cdots & \end{array}$$

but if they did both flip simultaneously and their neighbors remained unchanged, yielding

$$
\begin{array}{cccccc}
 & \cdots & [\uparrow] & [\uparrow] & \cdots & \\
\cdots & [\uparrow] & [\uparrow] & [\uparrow] & [\downarrow] & \cdots \\
 & \cdots & [\downarrow] & [\uparrow] & \cdots &
\end{array}
$$

then each of their bond energies will have changed from 0 to -2 with a resulting change in the total energy. To avoid this we place the spin sites on a black/white checkerboard grid and we apply the flip rule by alternating between the black and white sites. This circumvents the aforementioned difficulty and results in a reversible cellular automaton. This procedure can still be construed as simultaneous updating of all cells by introducing a spatial parameter that alternates at each time step between the values 1 on the black cells, 0 on the white cells, and 0 on the black cells, 1 on the white cells. Each cell with the parameter value 1 will have the flip rule applied, and if the cell value is 0, then the cell state is to remain unchanged.

More specifically, the total energy is computed for any initial configuration by first taking the total bond energies for all the black cells of the lattice, allowing all spins on black cells to flip that are energy preserving, and then computing the energies of the spins on all white cells and flipping all of those spins that are energy preserving. The total energy is computed by taking one-half the sum of the energies of the spins on black and white cells, respectively.

For a typical simulation, with proportions of spins up (and down) whose energy levels are greater than the critical value, the system evolves to a more or less stable configuration of half the spins up and half the spins down whose magnetization is zero as in Fig. 5.31.

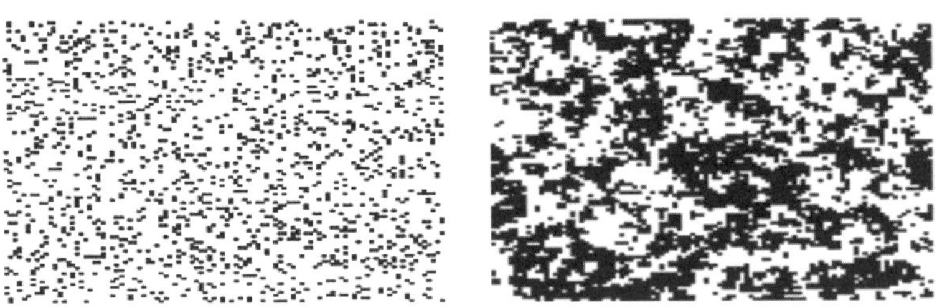

Figure 5.31 An Ising spin simulation with an initial random configuration of 15% spins up (black) and after 1000 iterations. This shows the system evolving to the equilibrium state of 50% up spins and 50% down spins with a magnetization of zero.

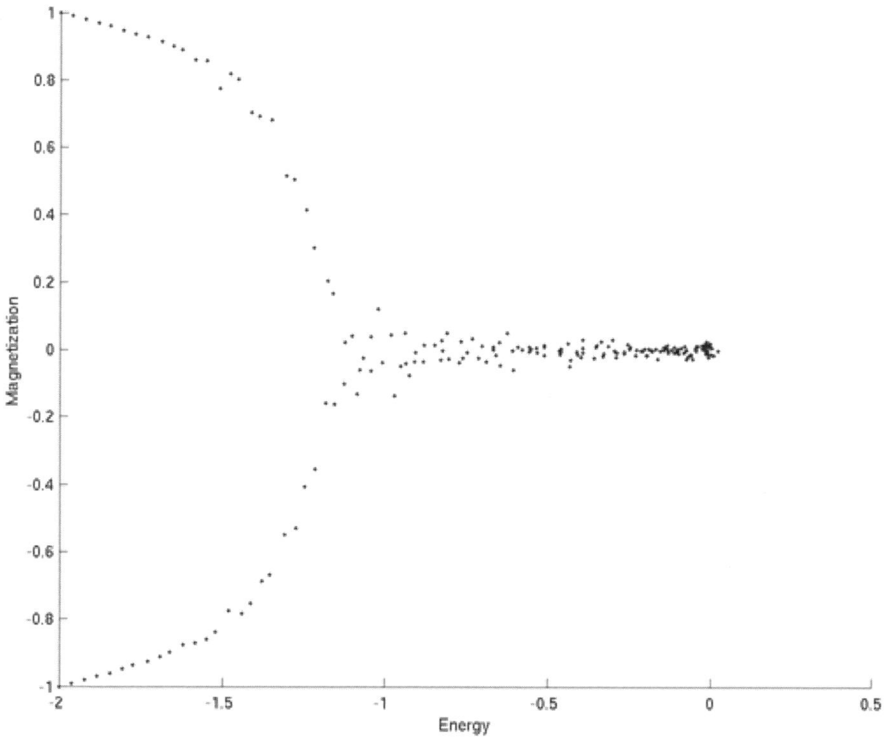

Figure 5.32 In this depiction of *magnetization* vs. *energy*, we have considered on a checkerboard lattice a range of random initial sites with the proportion of up spins ranging from 0% to 100%, respectively. The spins that are energy preserving on black and white cells are flipped until the magnetization stabilizes and that value is then plotted for the given energy value. Note that the graph is very close to the ideal depicted previously.

Generally, there is a small degree of fluctuation of the magnetization over subsequent iterations that preserve energy because the system is finite.

Moreover, if we also plot values of energy *versus* magnetization for the equilibrium configuration after the system has settled down, and do this over a wide energy range, we find a relationship close to the idealized one (Fig. 5.32). Just below the critical value, the equilibrium magnetization can fluctuate rapidly between the higher and lower states, but the overall result looks much like the ideal graph.

Another more sophisticated cellular automata model of Ising spin dynamics has been developed by Creutz [1986], but the algorithm also conserves constant total energy and updates cell spins in a checkerboard fashion. There are many important variations on the Lens-Ising theme that have emerged over the years, with names like, "Heisenberg model," "vector Potts model," and "XY model."

5.5.6 Steady-State Heat Flow

By allowing the cells of our lattice to have any values whatsoever, we can obtain quantitative information on a whole host of problems in physics. One such is the determination of the steady-state temperature (i.e., not changing with time) of the interior of a homogeneous body when only the surface temperature is known. The temperature at each point of the interior is governed by a famous partial differential equation known as the Laplace equation:

$$\frac{\partial^2 u}{\partial x^2} + \frac{\partial^2 u}{\partial y^2} + \frac{\partial^2 u}{\partial z^2} = 0,$$

where $u(x, y, z)$ denotes the temperature at each point (x, y, z). This equation is also satisfied by electrostatic and gravitational potentials and is thus one of the fundamental equations governing how the world works.

Let us consider a solid sphere of some homogeneous material and suppose that the surface of the sphere is held at a specified temperature which may vary continuously from point to point on the surface. The interior will in time reach an equilibrium steady-state temperature which we now seek to determine. The salient feature about this situation is that the temperature at the center of the sphere should be exactly the *average* of all the values on the surface of the sphere. This is so because each point on the surface is equidistant from the center and thus influences the temperature there exactly to the same degree (no pun intended); see Fig. 5.33.

Mother Nature knows this and makes this obvious choice for the temperature at the center. This feature of the steady-state temperature (or electrostatic and gravitational potentials for that matter) is called the *mean-value property.* The same property holds if we consider a thin

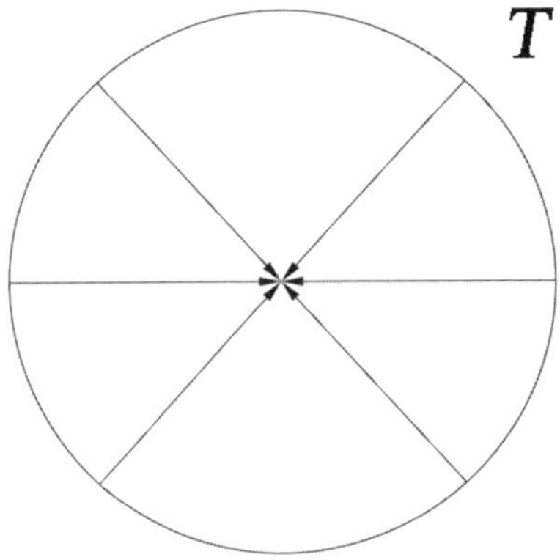

Figure 5.33 A circular region with a continuous temperature distribution T on the boundary. Since every point on the boundary is equidistant from the center, the (steady-state) temperature at the center will be the average of all the values of T.

circular disk and put a suitable temperature distribution on its circumference. The steady-state temperature at the center of the disk will be the average of the temperature values on the circumference. For this important property of the mean value to hold, the boundary temperature distribution should essentially be continuous but it can have a finite number of jumps or *discontinuities*. These are unimportant and can be dealt with as well. We will employ such a boundary temperature with a few discontinuities in the example below.

Suppose that we wish to measure the temperature inside a rectangular region that is twice as long as it is wide and say, for the sake of simplicity, this is just a two-dimensional figure. Let us put a temperature distribution of 100°F on one long side and maintain 0°F on the three other sides, as in the Fig. 5.34.

Let us try to find the steady-state temperature at the *center* of the rectangle. We fill up the rectangular region with a lattice of square cells and can cheat a trifle here by putting in 201 square cells along the long sides of the rectangle and 101 square cells along the shorter sides, thus giving us a cell that surrounds the point in the center. The ratio of length to width is not exactly 2-to-1, but certainly near enough. Or as mathematicians are wont to do, just change the problem to a rectangle that is 201 units by 101 units! The cells bordering one long side take

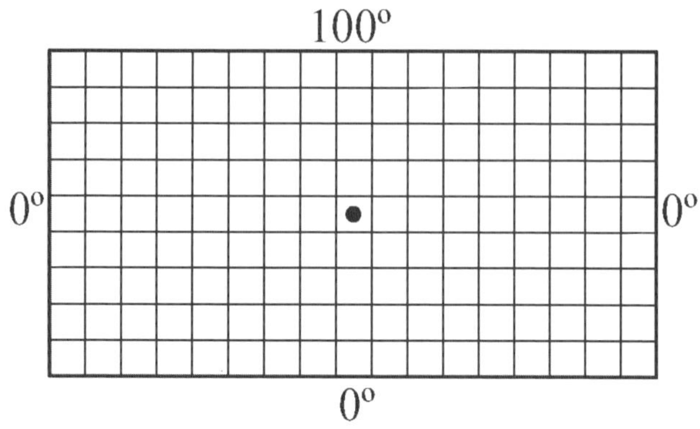

Figure 5.34 The rectangular region with length twice as long as it is wide (idealized here) and having a temperature distribution as indicated on its boundary. How warm will it be at the center of the figure?

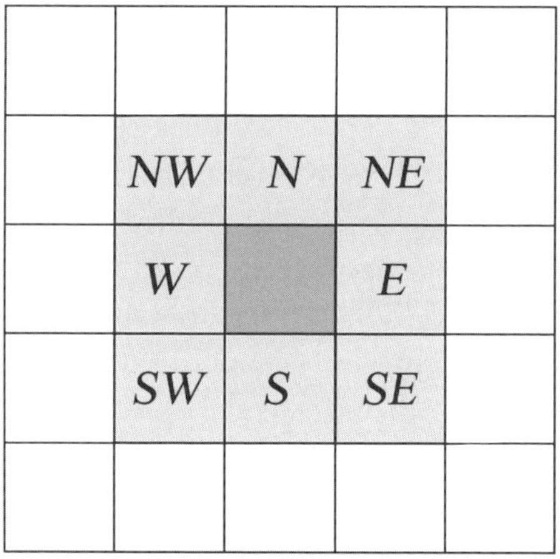

Figure 5.35 The eight neighbors taken in the mean-value temperature calculation of the central cell.

the fixed boundary value of 100°F and the cells along the other sides maintain the fixed value 0°F, as in Fig. 5.34.

All we need now is the transition rule which will follow naturally from the mean-value property. Taking a Moore neighborhood for each cell, the eight neighbors around the central cell will be a sufficient approximation to a circular neighborhood (Fig. 5.35).

By the mean-value property, each central cell should be the average of its neighbors and so this is how each central cell will be updated at the next time step. However, the cell values at the corners of the Moore neighborhood, namely, NW, NE, SW, SE, do not have the same influence as the cell values N, S, E, W that are directly facing the central one, and so we will take their value to be one-fourth that of the former ones. This gives us the formula for the transition rule:

$$T = \frac{4(N + S + E + W) + (NW + NE + SW + SE)}{20},$$

where T is the temperature at the next time step of the central cell. We can check this is correct by putting a temperature of 1 in each of the eight neighborhood cells. Then the value at the center will be $T = [4(1+1+1+1)+(1+1+1+1)]/20 = 1$, the desired "averaged" value. This is exactly the same kind of average that was taken in the bacterial colony model (Section 5.4.3).

Running the cellular automaton (~18,000 time steps) and starting with a zero initial value for all cells except the ones on the top row, which are maintained throughout at 100°F, as well as maintaining the other boundary values at 0°F, we find that the value of the central cell converges to $T = 44.4403$°F. Fig. C.18 (see color section) shows a color rendering of the steady-state temperature distribution in the whole room since the temperature in each cell is calculated in the CA process. Thus we have actually used in the entire procedure a finite number of cell values (states), but we did not *a priori* specify each of them, and so the number of cell states must be considered potentially infinite. While we have not actually found the temperature at the central point of our region, but rather a value for the whole of the central cell, this is typical of what can actually be measured in practice.

There is another rather unusual way in which to determine the value of the temperature at the center of our rectangle. In 1944, the Japanese mathematician Shizuo Kakutani discovered that for a problem such as this one the temperature at any interior point p of the rectangle would be equal to the probability (times 100) of a "random walk" that started at p first hitting the side having the 100°F, that is, the top side.

What is a random walk? Basically, it is a path consisting of very small unit-length steps such that each step is taken in a random direction, like the path a drunkard would take after a few too many (Fig. 5.36).

So, if we start a particle at the center of the rectangle and take a sequence of small steps each time in a random direction, eventually we will hit one of the sides (it is a mathematical fact that we will not wander

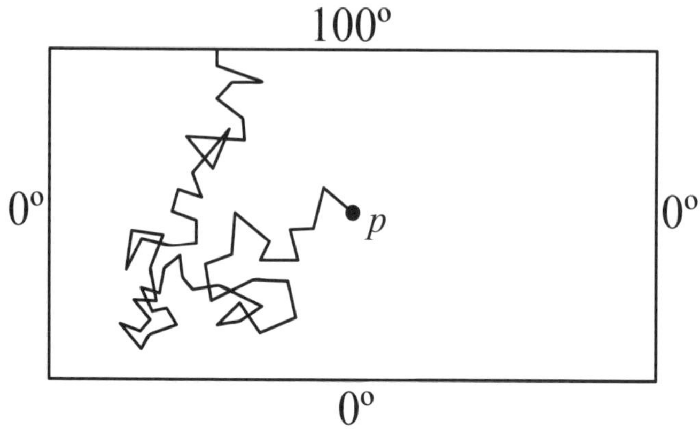

Figure 5.36 Random walk simulation inside a rectangle in order to determine the temperature at the starting point at the center.

around the interior of the rectangle forever). If we first hit the side with the 100°F, we stop the particle and keep a tally of this and repeat the random walk of the particle from the center again and again after each time we hit a side. The number of hits on the 100°F side divided by the total number of random walks taken gives us the desired probability of hitting the 100°F side first. This little algorithm is easily implemented on a computer and the simulation was run a total of 10 times taking 2000 random walks each time, with the average of these 10 values at the center of the rectangle being $T = 44.3500$°F, which is in very good agreement with our method using a cellular automaton. Of course, each time we ran the random walk simulation we obtained a slightly different value because each individual random walk does indeed have a random outcome, which will not be the case in the cellular automaton approach.

Another more analytical mathematical technique is available to solve this problem, called the *method of separation of variables.* Without going into the technical mathematical details, the answer it gives for the temperature $u(x, y)$ at any interior point (x, y) is in the form of an infinite series,

$$u(x,y) = \frac{400}{\pi} \sum_{n=1}^{\infty} \frac{\sinh((2n-1)\pi y/a)}{(2n-1)\sinh((2n-1)\pi b/a)} \sin\left(\frac{(2n-1)\pi x}{a}\right),$$

where a is the length of the rectangle and b is the height of the rectangle.

If we put in the coordinates of the center of the rectangle, namely $(100.5, 50.5)$, with $a = 201, b = 101$, and take a suitably large but finite number of terms of the series in order to calculate the sum (so even in the analytic case we are forced to make a finite approximation), we find that $u(100.5, 50.5) = 44.4258$°F. This value is the most accurate, but the others are in excellent agreement with it. The value obtained by using cellular automata could have been improved further by taking the cell size even smaller, since the temperature we obtained is that for the entire central cell and we are really interested in the value at the central point. On the other hand, it should be borne in mind that the notion of a *point* is merely a mathematical idealization, whereas actual (mean) temperature measurements must be taken over a region of space, thus lending credence to the CA approach. In fact, we must be careful not to take the sampled region *too* small, lest the temperature readings start breaking up (see Toffoli [1984]).

Since points are an ideal, perhaps, we should view the whole world as discrete.

5.5.7 The Digital Universe of Konrad Zuse and Edward Fredkin

Your theory is crazy, but it's not crazy enough to be true.
— *Physicist Werner Heisenberg to Wolfgang Pauli*

Many entities in the physical world are indeed discrete. Even though a table appears like a solid object, we know that it is made up of atoms which themselves are composed of discrete subatomic particles. Light is another entity that can be considered to be made up of discrete photons, and even the angular momentum of electrons in the Bohr model of an atom is a discrete quantity.

It is now thought by many scientists that the fabric of space-time is also discrete rather than smooth and continuous. Although this at first seems highly contradictory to experience, much of quantum mechanics, which is the physics of the subatomic world, has no counterparts in everyday experience. In quantum mechanics, the smallest components of space-time are the Planck length $\lambda_p = \sqrt{hG/c^3} \approx 10^{-33}$ cm (where h is Planck's constant, G is the universal gravitational constant, and c is the velocity of light), and Planck time $t_p \approx 10^{-43}$ seconds. The two are related in that t_p is the time it takes for a photon of light to travel a distance of λ_p, so that, $t_p = \lambda_p/c$. These are the smallest measurements of length/time that have any meaning in physics as we know it. So it is quite possible that at these scales the physical world looks granular rather than continuous and that it evolves in a steplike process, much like the still frames in a movie film. Some work on a discrete quantum mechanics has been initiated by Succi [2002].

The notion that the laws of physics are discrete and the universe evolves as the result of deterministic computation, such as a cellular automaton, dates back to pioneering German computer engineer Konrad Zuse [1969]. This proposal, referred to as "Zuse's thesis," has led to the development of the field of *digital physics*. One interesting consequence of the Zuse thesis is that entropy does not increase. Zuse is also credited for developing and building the first fully functional program-controlled digital computer, the Z3, in 1941. He was clearly a man ahead of his time. For more on Zuse, digital physics, and other ideas germane to this book, see the eclectic website of Jürgen Schmidhuber, `http://www.idsia.ch/~juergen/`.

Zuse's thesis was elaborated upon by computer scientist Edward Fredkin [1990, 1993], who proposed the related notion of finite Nature which is "the assumption that, at some scale, space and time are discrete and that the number of possible states of every finite volume of space-time is finite." Let us explore this idea further and see what are some of its consequences.

The scale of the CA cells is somewhere between the Planck length and the Fermi distance of 10^{-13} cm, which is about the size of a proton or neutron. In fact, the exact scale is not so important for many of the consequences of the theory. Each cell has the usual three-dimensional space coordinates x, y, z, as well as a time coordinate t in an absolute fixed reference lattice of cells.

One important consequence of the assumption that all entities in Nature are discrete is that each cell contains only a *finite* amount of information and thus can be coded as a finite string of integers representing a particular state. Then in order for the discrete universe to evolve with (discrete) time, it must be acted upon by a *fundamental informational process* that takes the present state of the cells, and probably their past state as well, in order to determine their future, in other words, a second order system. In the resulting model of physics that Fredkin calls digital mechanics, the convention is adopted that a particle contains such information as its momentum, energy, spin, and rest mass in digital form.

A feature such as the rectilinear motion of a particle would be achieved by the sequential updating of the particle's position with respect to the reference lattice. Such a preferred reference coordinate system is indeed essential for the description of motion in an informational based system. But Fredkin states that these coordinates are most likely implicit and contained in the bits that represents a particle's other properties, rather than explicit in the fabric of the entire universe.

Of course, there is no experimental evidence to support the existence of such a universal reference frame, although according to Fredkin, the cosmic background radiation is one such example. The famous Michelson-Morley experiment of 1887 was the first of many experiments that explicitly failed to detect Earth's movement through an "aether wind" that was the supposed basis for such a reference frame. However, the existence of such an absolute reference frame does not contradict any of the laws of physics or the theory of relativity.

Support for a notion of this kind comes from the Nobel Prize winning physicist, Richard Feynman (The Character of Physical Law, *Messenger Lectures*, Cornell University, 1964, italics added):

> It has always bothered me that, according to the laws as we understand them today, it takes a computing machine an infinite number of logical operations to figure out what goes on in no matter how tiny a region of space, and no matter how tiny a region of time. How can all that be going on in that tiny space? Why should it take an infinite amount of logic to figure out what one tiny piece of space/time is going to do? So I have often made the hypothesis that ultimately physics will not require a mathematical statement, that in the end the machinery will be revealed, and *the laws will turn out to be simple, like the chequer board with all its apparent complexities.*

Albert Einstein was perhaps not thinking about chequer boards or cellular automata (the latter not having been invented), when he expressed a similar view "that the principle of the universe will be beautiful and simple."

The fundamental process beating at the heart of digital mechanics, Fredkin views as one capable of universal computation, and the reason is that the laws of physics have allowed us to build computers that are capable of universal computation and hence the fundamental process should be capable of this as well. Moreover, the fundamental process is a reversible cellular automaton (because the laws of physics are) and so the universe is run by a reversible, universal cellular automaton, or RUCA. As Fredkin claims, "What cannot be programmed cannot be physics."

Reversibility implies there is a conservation law: *Information is conserved.* We have encountered this aspect of reversible cellular automata before and seen that reversibility allows one to retrieve the initial conditions no matter how complex the subsequent evolution has been (see Section 3.7). Information is lost in an irreversible system when two distinct configurations have the same successor so that it is not possible to choose which predecessor to return to. (Another of Fredkin's interesting achievements was the creation with Tommaso Toffoli [1982] of a billiard ball computer that allowed complete reversibility of computation, which is not normally a feature of even ordinary computers.)

One essential feature of quantum mechanics is randomness. Yet the randomness generated by a computer is deterministic and will be the same each time it is generated. But here we can invoke Wolfram's notion of computational irreducibility in that there is no way to predict the outcome of some computations and the only way to know the future is to compute it. This notion is embodied in the Fredkin statement, "In general, physics is computing the future as fast as it can."

Concepts such as conservation of energy and momentum for particles that interact, are, in the Fredkin world, consequence of the sharing of the information that the particles possess about their energy and momentum. Since the energy and momentum information cannot be lost, it is just reshuffled among the particles and hence conserved.

So far so good, and many scientists might agree that in some sense the universe acts *as if it were* a cellular automaton. However, in Fredkin's view, it *is* a cellular automaton. This means that there is some computer, *the engine*, that runs the fundamental processes that run physics and that this engine exists outside our universe.

Now the theory is beginning to sound a bit crazy, but could it be crazy enough to be true?

CHAPTER 6

COMPLEXITY

Order vast, order ordained, order for free.
— *Stuart Kauffman* – At Home in the Universe

The physicist's problem is the problem of ultimate origins and ultimate natural laws. The biologist's problem is the problem of complexity.
— *Richard Dawkins* – The Blind Watchmaker

Life is more than just clockwork, even though it is just clockwork.
— *Steve Grand* – Creation: Life and How to Make It

We have seen that with cellular automata one can sometimes achieve self-organization whereby a higher level order is created from the local interaction of autonomous agents obeying simple rules. This type of behavior of a system of many interacting units that engenders some self-organizing global order that is not present at the lower level is part of the notion of *complexity*. The self-organization is an *emergent* property in that it emerges spontaneously from simple local interactions with the result that in some sense the whole is greater than the sum of its parts.

This is contrary to the conventionally held reductionist view in science that the whole can be understood only through a thorough understanding of each of its parts. Who would have expected that the power of a desktop computer lay hidden in the transition function for elementary Rule 110? Could one ever have anticipated the vast diversity of lifeforms that arose in the Game of Life or their complex interactions that produced new lifeforms simply from knowing the rules of the game? Or the completely ordered structure that arose from the mists of the seemingly random conditions of Rule 94R in Fig. 3.24? None of this order was imposed by recipe or design (which is a completely different way to achieve a complicated structure such as in the blueprints for a nuclear submarine) but was rather an invisibly preordained consequence of the rules of local interaction, or as Kauffman says, "order for free." It is the "ghost in the machine" if you will.

Moreover, according to Camazine *et al.* [2003], "Complex systems ... may show diverse responses that are often sensitively dependent on both the initial state of the system and nonlinear interactions among its components. Since these nonlinear interactions involve amplification or cooperativity, complex behaviors may emerge even though the system components may be similar and follow simple rules. Complexity in a system does not require complicated components or numerous complicated rules of interaction." Indeed, this has been in evidence throughout all our investigations thus far.

Self-organization can be either *intrinsic*, as in the cellular automata examples examined thus far, or it can be *extrinsic* and come about by some external stimulus, such as in the self-aggregation of the slime mold discussed in the sequel.

One problem that needs to be faced in dealing with complexity is how to define and measure it. Many people have an intuitive understanding of the concept and would agree, for example, that the human brain is more complex than that of a rat, but there is no formally accepted definition. In fact, it means somewhat different things in different fields. Ilachinski in his book on cellular automata [2002] discusses eight measures of complexity from varying perspectives under the headings algorithmic complexity, computational complexity, graph complexity, hierarchical complexity, Shannon's information, logical depth, simplicial complexity, and thermodynamic depth. All of them are of some value in the appropriate context, but all have some drawbacks as well.

In computer science, complexity is related to the theoretical limits of computation itself. In biology, Dawkins [2004] gives the example (borrowing ideas from information theory) of comparing the complexity of a lobster to a millipede by writing a book describing a lobster and writing another one "down to the same level of detail" as the millipede. The lobster book is longer and so the lobster is a more complex organism. But Shannon information obeys an additive property, so this cannot be the whole story of complexity because we should not be able to generate more complex systems by simply adjoining multiple copies of simpler ones. Indeed, "The millipede book would consist of one chapter describing a typical segment, followed by the phrase, 'Repeat N times', where N is the number of segments" (Dawkins [2004]). Actually, the whole notion of biological complexity very much exercises the minds of evolutionary biologists such as Dawkins and others, because highly complex systems like the human brain have already evolved over millions of years from simpler forms and are not just theoretical constructs. Indeed, Stuart Kauffman goes so far as to say that the complexity found in Nature is not only the result of natural selection but the result of self-organization as well.

Likewise, the notion of emergence is somewhat ill-defined. "None of us has a solid grasp of emergence, far less a full definition" (John Holland, in Lewin [1999]). Notwithstanding, see his book on the subject, Holland [2000]. So, in the sequel, we will content ourselves by illustrating the concepts of complexity and emergence with a number of examples. Of course, we have already seen many such examples of both concepts throughout the text. Popular accounts of the subject of emergence and complexity from a variety of perspectives can be found in Johnson [2002], Lewin [1999], and Waldrop [1992].

6.1 MIND OVER MATTER

Dubito ergo cogito: cogito ergo sum — I doubt, therefore I think: I think therefore I am. Descartes' venerable dictum. We are conscious of our own existence. But wherein lies the "I," where is the "you," dear reader inside your head? I am trying to carry on a discourse with you, do you read me? If I dissect your brain bit by bit, will I eventually find you? I will have to wade through some 100 billion neurons and it would certainly not be an easy task as each neuron can be connected to thousands or tens of thousands of others. And I don't even know what I'm looking for! Writing from some time in the future, Paul Broks of the University of Plymouth reminisces that "consciousness" went the way of phlogiston — the hypothetical substance once conjured up by scientists to explain fire (see his thought provoking article in *New Scientist*, November 18, 2006, pp. 56–61). Perhaps you are just an illusion, merely a bundle of thoughts and experiences, or as Broks eloquently puts it, "The enchanted loom of the brain does not require a weaver." In other words, nobody's home.

So, before we turn our attention to the rather difficult and contentious issue of consciousness, let us first consider the business end of the brain — neurons (Fig. 6.1). Since the brain's many functions are brought about by the interaction of interconnecting neurons, we will begin with a very simplified but heuristic model of this interaction. The seminal McCulloch-Pitts model [1943] maintains that the state of the ith neuron at the next time step is given by

$$c_i(t+1) = H\left(\sum_j w_{ij}c_j(t) - \theta_i\right),$$

where H is the Heaviside step function, w_{ij} are weight values given to the coupling strength between (presynaptic) neuron j and (postsynaptic) neuron i, with $c_i(t)$ taking on the values 0 and 1 and θ_i a threshold value for the i th neuron. Thus if the weighted sum of all incoming signals to a neuron achieves a threshold value, the neuron fires, taking state value 1. If the threshold is not reached, the state value of the neuron is 0. This is reminiscent of the earlier CA example, vote, in which the weights w_{ij} all had the value 1 and the threshold was $\theta = 5$. The rather difficult paper by McCulloch and Pitts containing this model spawned a large amount of research into these artificial neurons that continues to the present. Frank Rosenblatt [1958] initiated work on *perceptrons* that in the simplest case of a single neuron uses a feedback mechanism (*back-propagation*) to compare the neuron's output with a desired output and, if the two are not the same, adjusts the weights in a suitable manner based on an error correcting factor at each time step. For a certain class of problems, known as *linearly separable*, the solution will be found by the (single-layer) perceptron in a finite number of steps. Investigated more fully by Minsky and Papert [1969] in their now classic text, *Perceptrons: An Essay in Computational Geometry,*

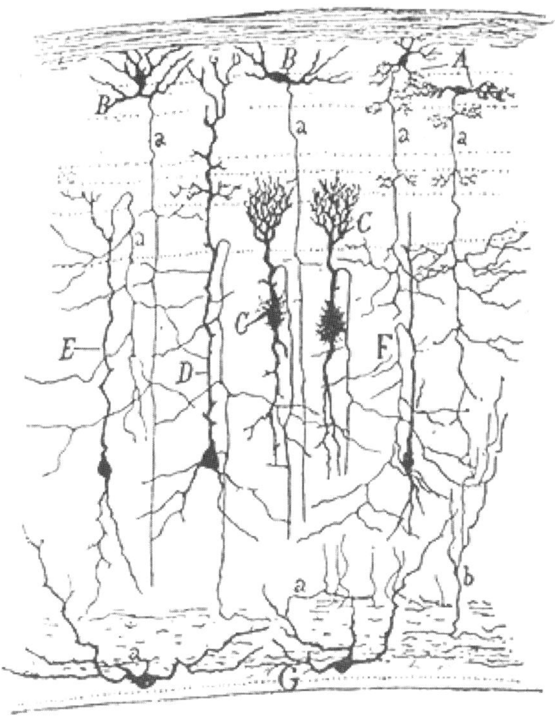

Figure 6.1 Neurons from the *optic tectum* of a sparrow drawn by one of the founders of neuroscience and Nobel Prize winner Santiago Ramon y Cajal (1852 – 1934). A variety of neuron types are illustrated. In (B) the *soma* is the irregular mass containing the nucleus of the cell. Above the soma are branchlike dendrites that pick up the incoming signals. The fiber extending downward from the soma marked (a) is the *axon* that carries the output signal. Note the very bushy dendrites in (C). The staining technique used at the time does not permit the visualization of all the neurons present, and thus many dendrites do not appear connected to anything. In fact, there are numerous other neurons present but they are not visible.

the perceptron's shortcomings were fully disclosed. For instance, a (single-layer) perceptron is incapable of implementing the truth table of the EXCLUSIVE OR. However, a multi-layered approach has revived the fortunes of the perceptron since the early 1980s, and happily the EXCLUSIVE OR can be dealt with.

An assemblage of interconnected artificial neurons is called an *artificial neural network* and a model capable of learning was developed by

Hopfield [1982] (see the details of how it works in Ilachinski [2002]). Because such networks can be trained by an inductive learning process, artificial neural networks have been very useful in a variety of fields such as pattern recognition, finding data trends, medical diagnosis, various business applications, of course neuroscience, and many others, including spam filtering! One only has to check the proceedings of the 12th European Symposium on Artificial Neural Networks 2004 to see the diversity of current research: `http://www.informatik.uni-trier.de/~ley/db/conf/esann/esann2004.html`.

Neural networks have also been combined with a cell structure in a hybrid called *cellular neural networks* (Chua and Yang [1988 a,b]).

On the biological front, two-dimensional neural networks consisting of cultured cortical material extracted from one-day-old rats has been studied *in vitro* by E. Hulata *et al.* [2005]. Using microelectrodes, spontaneous activity consisting of synchronized bursting events of rapid firing of most neurons was detected and found to exhibit complex behavior.

The classical McCulloch-Pitts model uses an "all-or-nothing" threshold simulation of neuron firing with fixed weight factors, which is not actually the case for real neurons but is useful nonetheless for many simulation purposes. For example, these features allow one to simulate the basic logic gates that we have seen in Chapter 1. Actually, we'll make a slight modification to the Heaviside function and write

$$c_i(t+1) = \text{sgn}\left(\sum_j w_{ij}c_j(t) - \theta\right),$$

where sgn is the sign function that takes the values

$$\text{sgn}(x) = \begin{cases} 1 & \text{if } x \geq 0, \\ -1 & \text{if } x < 0, \end{cases}$$

and $c_i(t)$ takes on the values $+1$ and -1. With this change, we can implement the three logic gates AND, OR, and NOT in the McCulloch-Pitts model. What is more, the McCulloch-Pitts model demonstrated that a neural network composed of neurons firing synchronously at fixed time steps of this type can emulate a universal Turing machine and hence is capable of universal computation.

Fig. 6.2 is a schematic of a McCulloch-Pitts AND gate. The sum of the two weighted inputs exceeds the threshold only when they are both $+1$, giving an output of $+1$, and -1 otherwise.

If one or both of the inputs is $+1$ in the OR gate (Fig. 6.3), then the output is also $+1$. Otherwise the output is -1.

The input of $+1$ or -1 is reversed by the NOT gate (Fig. 6.4).

AND

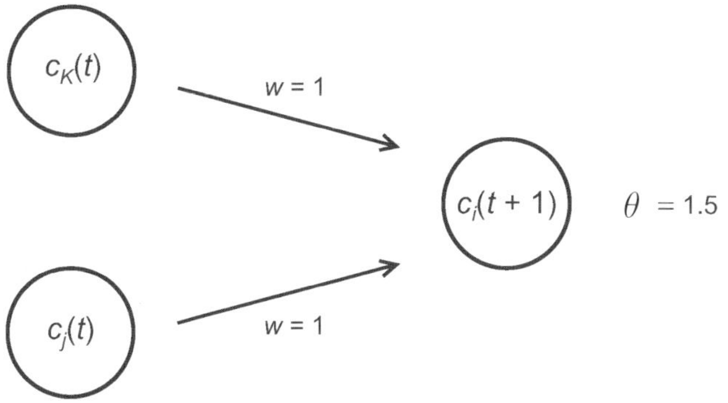

Figure 6.2 The AND gate requires the two inputs to both be $+1$ in order to exceed the threshold $\theta = 1.5$ enabling the ith neuron to fire at the next time step.

OR

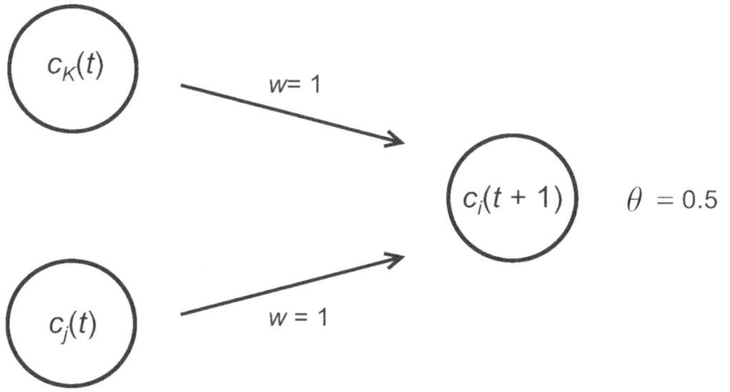

Figure 6.3 In the OR gate, if at least one of the inputs is $+1$, then the threshold will be exceeded and the ith neuron will fire at the next time step.

NOT

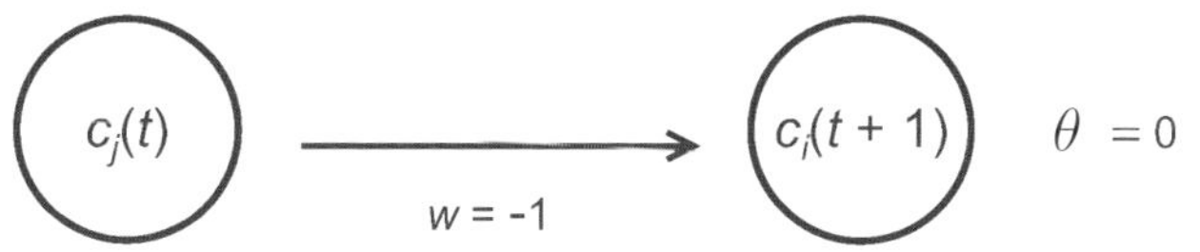

Figure 6.4 With the weighting factor $w = -1$, the output of the NOT gate is the reverse of the input.

This suggests, at least to some people, that the brain is nothing more than a highly sophisticated digital computer and that consciousness is an emergent property of complex neural activity. There is also the complementary notion that a digital computer can in principle perform all the functions of the human brain and itself achieve consciousness. These have been highly contentious scientific and philosophical issues. Nineteenth- century satirist Samuel Butler, while ruminating on his sheep station in New Zealand, was concerned about "the ultimate development of mechanical consciousness" and makes the following helpful suggestion, "Is it not safer to nip the mischief in the bud and to forbid them [machines] further progress?" Butler is certainly way ahead of his time. "But who can say that the vapour engine has not a kind of consciousness? Where does consciousness begin, and where end? Who can draw the line?" [*Erewhon*, 1872].

Sir Roger Penrose draws the line against this position throughout much of his book *The Emperor's New Mind* [1989]:

> "Is it not 'obvious' that mere computation cannot evoke pleasure or pain; that it cannot perceive poetry or the beauty of an evening sky or the magic of sounds; that it cannot hope or love or despair; that it cannot have a genuine autonomous purpose?...
>
> "Consciousness seems to me to be such an important phenomenon that I simply cannot believe that it is something just 'accidentally' conjured up by a complicated computation. It is the phenomenon whereby the universe's very existence is made known...
>
> "Yet beneath all this technicality is the feeling that it is indeed 'obvious' that the *conscious* mind cannot work like a computer, even though much of what is actually involved in mental activity might do so."

The reference to consciousness not being "accidentally" conjured up refers, of course, to the phenomenon of emergence. Rather, Penrose has worked out a theory of "quantum consciousness" in conjunction with anesthesiologist Stuart Hameroff, whereby consciousness is the result of effects that arise from quantum computation within *microtubules* that are inside the neurons themselves. Hameroff's website on quantum consciousness can be found by collapsing a few wave functions at `http://www.quantumconsciousness.org/`.

On the other hand, people like computer scientist Marvin Minsky take a completely opposite view:

> "We humans do not possess much consciousness. That is, we have very little natural ability to sense what happens within and outside ourselves...
>
> "In short, much of what is commonly attributed to consciousness is mythical — and this may in part be what has led people to think that the problem of consciousness is so very hard. My view is quite the opposite: that some machines are already potentially more conscious than are people, and that further enhancements would be relatively easy to make."

Minsky, in his book *Society of Mind* [1988], provides the following thought experiment. Imagine that all the neurons in your brain have been replaced by computer chips that perform exactly the same functions and are connected in the same fashion to other chips as the neurons. "There isn't any reason to doubt that the substitute machine would think and feel the same kinds of thoughts and feelings that you do — since it embodies all the same processes and memories. Indeed, it would surely be disposed to declare, with all your own intensity, that it is you." Perhaps, though, the devil is in the detail. Minsky's latest musings on consciousness, as well as the emotional mind, and a whole lot more can be found in his recent work, *The Emotion Machine* [2006].

CA researcher Norman Packard holds a similar view. In a conversation with Roger Lewin [1999] in his book *Complexity*, we have the following dialogue:

> "The simple evolutionary models of the sort I'm working with will eventually develop behavior rich enough that I'll see some kind of consciousness emerge."
>
> "You're saying that your computer model, a form of artificial life, will develop consciousness?"
>
> "I'm saying that the level of information processing in the system will evolve toward what we could call consciousness, that the organisms will reach a point where they will do information processing on their own, and become aware."
>
> "Artificial life, becoming aware of itself?"
>
> "Yes."

But there are others like philosopher John Searle who take up positions between Minsky and Penrose. Searle poses the *Chinese Room* thought experiment in which one is to imagine a room containing an English-only speaking person. This person has no knowledge of the Chinese language whatsoever, but what they do have is a handbook in which appropriate written responses in Chinese are given to questions that might be posed by a Chinese interrogator outside the room. When the interrogator slips a written question under the door, say, the person in the room simply looks up the appropriate response in their handbook and copies out the reply. The Chinese person outside the room could easily be convinced that the person inside the room understood Chinese, when in fact all they were doing was giving a symbolic response to a symbolic question. No actual understanding of the symbols was necessary and Searle claims that this is exactly what computers do — they manipulate symbols without being "conscious" of what they are doing. So even if computers are someday able to pass the Turing test and appear to have human intelligence, they still will not be conscious, if indeed even humans are!

One could perhaps argue that the ensemble of the room + person + book does have a clear understanding of Chinese, so Searle's line of argument is not necessarily the last word.

The debate over whether the human brain has consciousness or not, or a computer will some day be capable of consciousness or not, is likely to continue for some time to come. Let's not even worry about animals, although recent experiments have shown that Rhesus monkeys are capable of a certain degree of introspection — see *New Scientist*, December 16, 2006, pp. 28–31. There are others, such as Colin McGinn of Rutgers University who hold that an understanding of consciousness is beyond the scope of human senses in the same way that humans are unable to see in the infrared. We'll let Penrose have the last word. "Let's be honest. No one can really say what consciousness is." Still, it does make you wonder... .

Regarding complexity, the human brain is the most extraordinarily complex single entity in the solar system, maybe even in the universe. In the present context, perhaps that is all we can say for now.

6.2 RANDOM BOOLEAN NETWORKS

Rather than implement cellular automata in the usual array format, it is possible to consider a network whereby each cell becomes a node that is connected to various other nodes, not necessarily neighboring ones. The nodes can be considered in various states and these are updated synchronously at each time step by a local rule in accordance with the other nodes to which an individual node is connected. Because the network is finite in extent, the output tends to be similar to cellular automata of finite size, that is to say, periodic.

However, it is possible to enhance the dynamics of a finite network of nodes by allowing each node to operate under its own rule picked at random, which means that we have a generalization of a two-state cellular automaton known as a *random Boolean network* (*RBN*). The updating is still done synchronously and there are two allowable states for each node, 0 and 1 (off and on — Boolean). Each node has K inputs, randomly chosen from the other nodes, including itself (Fig. 6.5). Thus, there are 2^K different combinations of states and hence 2^{2^K} possible Boolean rules that can be formed. For example, if each node receives inputs from $K = 2$ other nodes, there are $2^{2^2} = 16$ possible Boolean functions to choose from for each node. We list them as

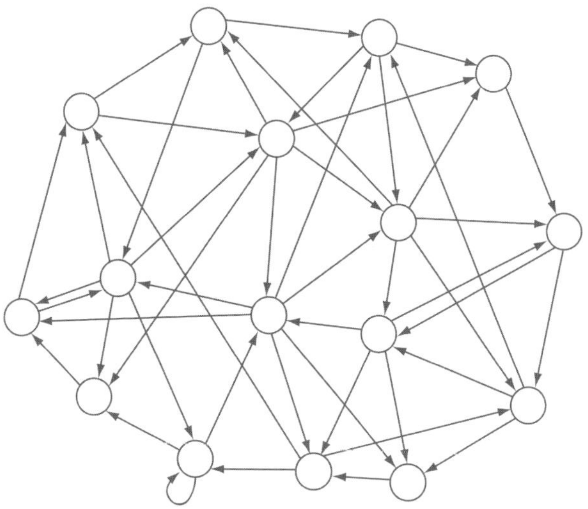

Figure 6.5 A random Boolean network with $N = 16$ nodes, each connected to $K = 3$ other nodes.

the following with O representing the output of the node receiving the given inputs I_1, I_2:

I_1	I_2	O	I_1	I_2	O	I_1	I_2	O	I_1	I_2	O
0	0	0	0	0	0	0	0	0	0	0	0
0	1	0	0	1	0	0	1	0	0	1	1
1	0	0	1	0	0	1	0	1	1	0	0
1	1	0	1	1	1	1	1	0	1	1	0

I_1	I_2	O	I_1	I_2	O	I_1	I_2	O	I_1	I_2	O
0	0	1	0	0	0	0	0	0	0	0	1
0	1	0	0	1	0	0	1	1	0	1	0
1	0	0	1	0	1	1	0	0	1	0	0
1	1	0	1	1	1	1	1	1	1	1	1

I_1	I_2	O	I_1	I_2	O	I_1	I_2	O	I_1	I_2	O
0	0	1	0	0	1	0	0	0	0	0	1
0	1	0	0	1	1	0	1	1	0	1	1
1	0	1	1	0	0	1	0	1	1	0	1
1	1	0	1	1	0	1	1	0	1	1	0

I_1	I_2	O	I_1	I_2	O	I_1	I_2	O	I_1	I_2	O
0	0	1	0	0	1	0	0	0	0	0	1
0	1	1	0	1	0	0	1	1	0	1	1
1	0	0	1	0	1	1	0	1	1	0	1
1	1	1	1	1	1	1	1	1	1	1	1

Some of these are previously encountered logic functions — the AND function is the second rule listed, the OR function is the second to last rule, and the EXCLUSIVE OR function is the eleventh rule.

Research on random Boolean networks was initiated by Stuart A. Kauffman [1969] and more fully discussed in his 1993 book, *The Origins of Order: Self-Organization and Selection in Evolution*, and in *Investigations* [2000] in order to study the emergence of order in biological systems such as gene networks. With two possible states of 0 and 1 for each node, if there are N nodes, then the number of possible states for the network as a whole to be in is 2^N and represents the size of the *state space*. Thus these models are also known as *NK* (or *Kauffman*) *networks*. It is helpful in this regard when thinking about the response of a network to consider the nodes as lightbulbs that are off when in state 0 and on when in state 1. No matter what the initial state of the network, it will eventually find itself in a state it has already reached and thus will repeat its behavior in a repeating *state cycle* much like an attractor of a dynamical system. The maximum possible length of a state cycle is of course the size of the state space itself, that is, 2^N. But even for a modest sized network of say $N = 200$ nodes, the state space is so large ($2^{200} \approx 10^{60}$) as to be effectively infinite in extent.

Kauffman spent years experimenting with Boolean networks, "counting the number of attractors, the lengths of attractors, the stability of attractors to perturbations and mutations, and so forth. Throwing the dice again, we can randomly wire another network with the same general characteristics and study its behavior. Sample by sample, we build up a portrait of a family of Boolean nets, and then we change the values of N and K and build up another portrait." It was found in the simplest case, when $K = 1$ where each node is only connected to one other, that the network exhibited very short state cycles, often of length 1 (*point attractor*). This sort of behavior characterizes the *ordered phase* of RBNs. A cycle of length 1 is analogous to a fixed-point attractor of a dynamical system.

At the extreme of network behavior, when $K = N$, which means that every node receives input from all others including itself, it was found that the length of the state cycles was on average proportional to $\sqrt{2^N}$, which is still virtually infinite even for $N = 200$. This means that one can wait for an eternity to observe the state cycle repeat itself in real time! In $K = N$ networks, if a single node is perturbed by having its state flipped, then the network's future evolution is also completely altered, thus exhibiting extreme sensitivity to initial conditions. These networks can be considered chaotic except for their modest number of state cycles. Similarly chaotic behavior was found in networks with $K \geq 3$ and this type of behavior characterizes the *chaotic phase* of RBNs.

When $K = 2$ (charcterizing the *critical phase*, see Fig. 6.6), the number and length of state cycles was long thought to look like $\sqrt{N}$ (Kauffman [1993]). This would have been nice because, even for such large networks with $N = 40,000$ and a state space having the outrageous number of $2^{40,000}$ possible states to roam around in, there would only be 200 state cycles! This would mean that the number of human genes (read "nodes") of somewhere between 25,000 and 40,000 (measuring the exact number is rather problematic) would roughly cor-

Figure 6.6 Once again, *Plus ça change...* Time steps run downward in these RBNs each with $N = 32$ (top row of nodes) and random initial states. Black represents a node that is on. Left to right represent trajectories in the ordered ($K = 1$), critical ($K = 2$), and chaotic ($K = 5$) phases. Note the similarities with cellular automata.

relate to the 254 cell types (read "state cycles") they give rise to, that is, the number of human cell types is comparable to the square root of the number of genes. This model was one of the motivations for Kauffman's work. However, two drawbacks with this scenario are that genes do not change their state synchronously and their connectivity (via regulating proteins) is rather higher than $K = 2$.

Furthermore, recent evidence has put the number of state cycles' dependence on $\sqrt{N}$ in dispute. The problem is that, as we saw for $N = 200$, the state space is simply so enormous that it is only possible to examine small portions of it, which of course will lead to problems of undersampling. More recent analyses have indicated a *linear* relationship between the number of attractors and the number of nodes N, as determined by Bilke and Sjunnesson [2001] and confirmed by others. Nevertheless, we have achieved an extraordinary degree of self-organization, or as Kauffman states, "stunning order... order for free."

The output for the $K = 2$ network depicted in Fig. 6.7 is given by the *lookup table*

1	2	3	$t \to t+1$	1	2	3
0	0	0		0	0	0
0	0	1		1	0	0
0	1	0		1	0	0
0	1	1		1	0	0
1	0	0		0	0	0
1	0	1		1	1	0
1	1	0		1	0	1
1	1	1		1	1	1

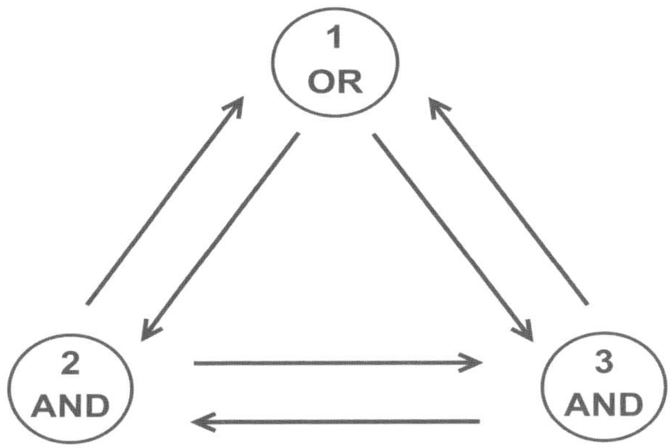

Figure 6.7 A schematic diagram of a RBN with $N = 3$ nodes and $K = 2$.

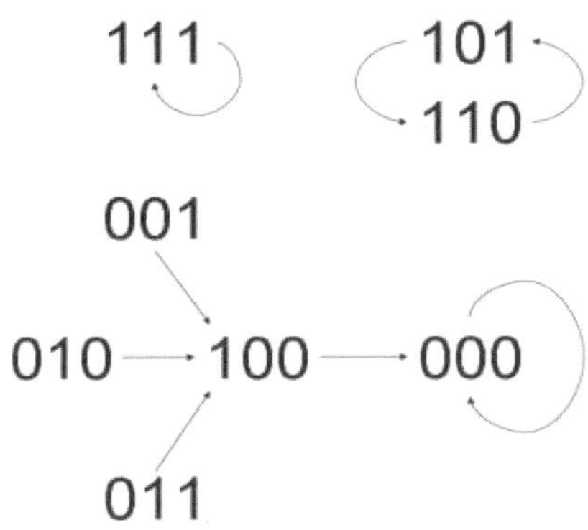

Figure 6.8 A graphical representation of the preceding lookup table and schematic diagram.

which gives the state transitions for the eight possible inputs for each of the three nodes at time t and $t + 1$, respectively. Node 1 is an OR gate, and nodes 2 and 3 act as AND gates in this example.

From Fig. 6.8 we note the point attractor $111 \rightleftarrows 111$, and the state cycle 2: $101 \rightleftarrows 110$. Analogous to dynamical systems, those states that flow into an attractor form its *basin of attraction* as observed in this example (see also Fig. 6.9 and Fig. C.19 in color section).

There is a Lyapunov exponent for RBNs introduced by Luque and Solé [2000] indicating phase transitions from ordered to chaotic ones, given by

$$\lambda = \log[2p(1 - p)K],$$

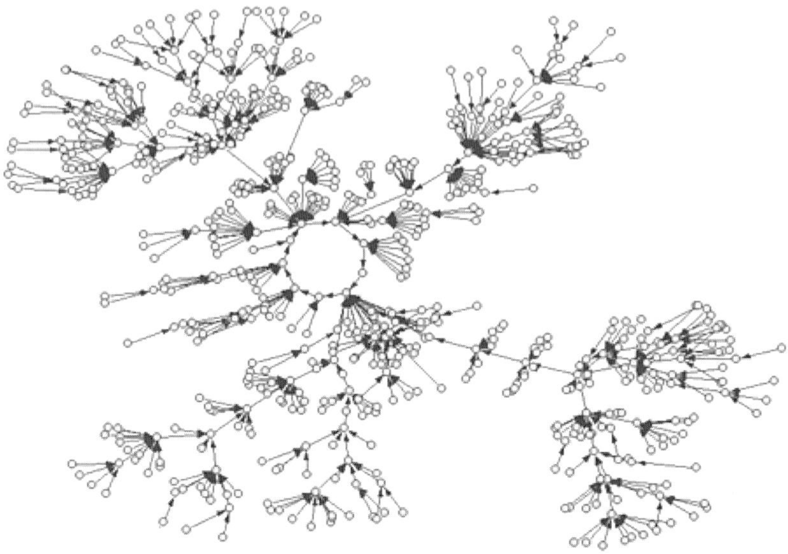

Figure 6.9 Depiction of the state space of an RBN, where each dot represents a different state of the network. The inner ring represents a state cycle of period 9, and the rest is its basin of attraction. States without any predecessors are known as Garden of Eden states, analogous to those found in cellular automata like the Game of Life.

where, as usual, K is the number of inputs and p is the proportion of 1's in the lookup table of the particular network. Here $\lambda < 0$ yields the ordered phase, $\lambda = 0$ the critical phase, and $\lambda > 0$ the chaotic phase. The relationship between K and p, order and chaos, is indicated in Fig. 6.10, where the *critical curve* dividing order from chaos is given by the equation

$$K = \frac{1}{2p(1-p)},$$

so that $p = 0.5$ corresponds to the critical phase $K = 2$.

There have been many attempts to alter the classical RBN model. Asynchronous RBNs have been introduced whereby a node is selected randomly and updated. As with asynchronous CA which lose much of the dynamics of their synchronous cousins, state cycles are lost in asynchronous RBNs as the system is no longer even deterministic. Multivalued networks have also been tried, the number of connections K can also be altered for different nodes, and in fact there is a veri-

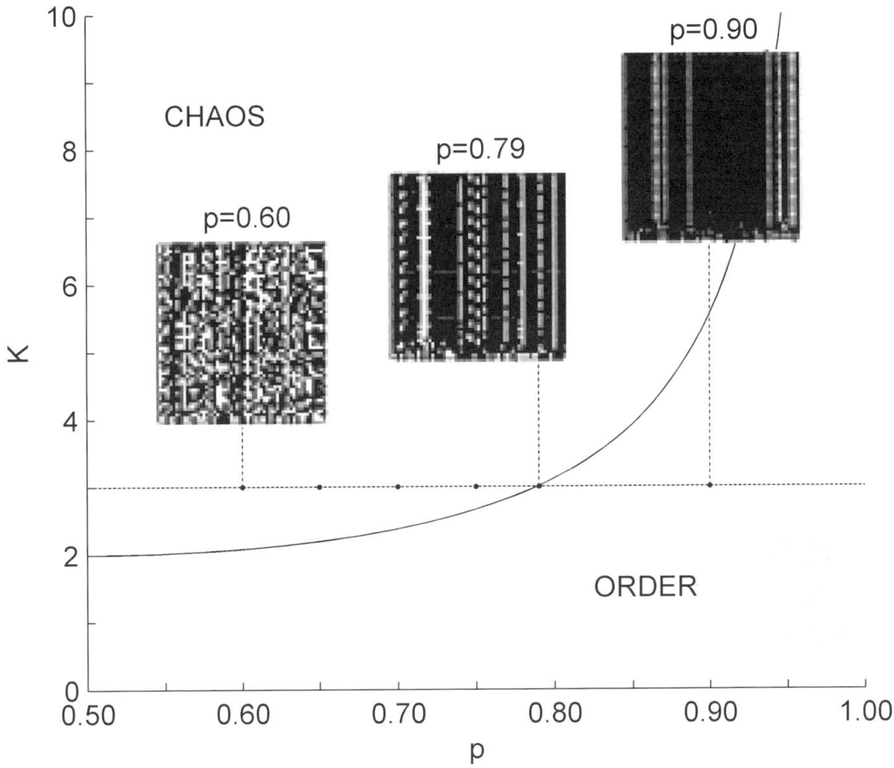

Figure 6.10 Phase space portrait with respect to the parameters K and p. The dashed line is shown for $K = 3$ and $p = 0.60$ yields the chaotic phase, $p = 0.90$ the ordered phase, and $p = 0.79$ the critical phase right on the transition curve, which can be considered at the edge of chaos. Ordered states for $K \geq 3$ are seen to exist for high p values.

table alphabet soup of variant RBNs, such as DARBNs, DGARBNs, GARBNs, and MxRBNs (see Gershenson [2004] for definitions).

Applications of RBNs are wideranging, from artificial life to real-life biology, computer science, the physics of chaos, and neural networks, among others. Still, a lot of work remains to be done in this field, and perhaps it will be by some of those who read this. For a comprehensive overview of the subject with an emphasis on Kauffman's original intent, see the paper by Kadanoff, Coppersmith, and Aldana [2002]. To conduct your own RBN explorations, first see the tutorial by Willadsen and Skellett at `http://www.itee.uq.edu.au/~kaiw/RBN/`, which is an introduction to the RBN toolbox in Matlab.

6.3 AUTONOMOUS AGENTS

The notion of an autonomous agent has taken on a life quite independent of cellular automata theory and is a major subject of research, such as in the study of complex adaptive systems. Often the differences between cellular automata and agent based models (or individual-based models — IBMs) are purely semantic. Over the years, numerous definitions of an autonomous agent have been proposed (see Franklin and Graesser [1996]), but a general one might be: *a computational system in a dynamic environment that senses this environment and acts autonomously within it.* Of course, an individual cell of a cellular automaton falls into this category, but some autonomous agent models are more general in nature than a cellular automaton. A notable example is the popular interactive computer simulation *The Sims* that allows the user to interact with a host of humanlike autonomous agents. Autonomous agents have also been used to model political conflicts (Ian Lustick's *Virtual Pakistan*, and *Middle East Polity*) as well as military ones (the Pentagon's *OneSAF*). Of course, all of these are just models, based on the rules built into them, so that the output is virtual and thus not necessarily real. On the other hand, Lustick has successfully tested *Virtual Pakistan* against real field data, like "What proportion of Pushtuns who live in cities are also part of the commercial elite?" even though this sort of data was not part of the program itself (*Popular Science* March, 2004).

In this regard, a particularly useful software development is *Swarm*, created by Christopher Langton and developed at the Santa Fe Institute in New Mexico. It provides the environment in which autonomous agents can be employed to model a whole host of collective behavior and has been used by scientists to study, among other things, rain forest ecosystems and nuclear fission chain reactions.

Swarm is intended to be a useful tool for researchers in a variety of disciplines. The basic architecture of *Swarm* is the simulation of collections of concurrently interacting agents: with this architecture, we can implement a large variety of agent-based models. For an excellent presentation of how self-organization pertains to swarms and its many applications to problems of optimization, see the monograph by Bonabeau, Dorigo, and Theraulaz [1999].

6.3.1 Honey Bees

One creature that does indeed swarm is the honey bee, a subject of much fascination (Camazine *et al.* [2003]):

> The challenge here is to understand how a colony-level (global) structure emerges from the activities of many individuals, each working more or less independently. In the building of these large biological superstructures, it is difficult to imagine that any individual in the group possesses a detailed blueprint or plan for the structure it is building. The structures are orders of magnitude larger than a single individual and their construction may span many individual lifetimes. It seems far more likely that each worker has only a local spatial and temporal perspective of the structure to which it contributes. Nonetheless, the overall construction proceeds in an orderly manner, as if some omniscient architect were carefully overseeing and guiding the process.

The authors then proceed to demonstrate how a characteristic pattern develops on the honeycomb due to a self-organizing process of the bees.

But let us look into the actual construction of the honeycomb. First of all, each cell of the honeycomb is hexagonal in shape (Fig. 6.11), which itself appears to be a self-organizing phenomenon: "There is a theory, originally propounded by Buffon and now revived, which assumes that the bees have not the least intention of constructing hexagons with a pyramidal base, but that their desire is merely to contrive round cells in the wax; only, that as their neighbours, and those at work on the opposite sides of the comb, are digging at the same moment and with the same intentions, the points where the cells meet must of necessity become hexagonal" (Maeterlinck [1901]).

Each cell of the honeycomb is about 12 mm long and the worker bees deposit royal jelly into them onto which the egg is laid. The honeycomb is double sided with the cells from each side meeting in the middle. But the base of each cell is not just flat across the bottom because a little depression or well is required to keep the royal jelly in place as well as the egg. The bees create this depression in the form of a three-sided pyramid. The pyramidal bases for each cell from one side of the honeycomb are interleaved with the pyramidal bases coming from the other side of the honeycomb so that it all fits snugly together.

But something even more profound happens in the construction of the honeycomb. The question one can ask is, "At what angles should the three planes of the pyramid meet if one wished to minimize the

Figure 6.11 The honeycomb consists of a double-sided hexagonal array of cells wih a pyramidal base at the bottom end of each cell.

amount of labor and material that goes into their construction?" This turns out to be a rather nontrivial mathematical question requiring a knowledge of calculus (Maeterlinck [1901]):

> It has been demonstrated that, by making the bottoms of the cells to consist of three planes meeting in a point, there is a saving of material and labour in no way inconsiderable. The bees, as if acquainted with these principles of solid geometry, follow them most accurately. It is a curious mathematical problem at what precise angle the three planes which compose the bottom of a cell ought to meet, in order to make the greatest possible saving, or the least expense of material and labour. This is one of the problems which belong to the higher parts of mathematics. It has accordingly been resolved by some mathematicians, particularly the ingenious Maclaurin, by a fluctionary calculation which is to be found in the Transactions of the Royal Society of London [1743]. He has determined precisely the angle required [70° 32′, 109° 28′], and he found, by the most exact mensuration the subject would admit, that it is the very angle in which the three planes at the bottom of the cell of a honeycomb do actually meet.

Maeterlinck then attempts to come to terms with the notion of complexity well before the concept was ever enunciated, "I myself do not believe that the bees indulge in these abstruse calculations; but, on the other hand, it seems equally impossible to me that such astounding results can be due to chance alone, or to the mere force of circumstance."

In a recent expository article by Peter Miller [2007] in *National Geographic*, certain principles of swarm intelligence exhibited by honey bees are deemed to be valuable lessons for human group decision making. We are still learning from the birds and the bees.

6.3.2 Slime Molds

Another model based on Turing's morphogenesis work was formulated by Evelyn Keller and Lee Segel [1970] to describe the unusual behavior of the species of slime mold, *Dictyostelium discoideum.* It consists of thousands of independent-living amoebae that feed off bacteria on the forest floor and normally live and move freely from one another. During periods of adverse conditions such as food scarcity, up to 100,000 cells aggregrate into a single organism (the *slug* stage) that moves about in search of food. At a later stage, the slug transforms itself into a *fruiting body* comprised of a globule of spores sitting atop a slender stalk. Under suitable conditions the spores are released and germinate to begin the life-cycle anew. How this oscillation between single organism – multiple organisms is achieved has been the subject of intense scientific investigation. Prior to Keller and Segel's work, the prevailing theory due to B. M. Shafer of Harvard was that so-called pacemaker cells of the slime mold would release a chemical signal, known as cyclic AMP, which would trigger other cells to do the same and so on throughout the dispersed community of cells. The cyclic AMP was a signal to re-form the aggregate. The only difficulty with this theory, since the release of cyclic AMP was well documented, was that all slime mold cells were much like any other and no pacemakers could be found.

According to the Keller-Segel version of the theory, the individual amoebae begin to aggregrate on their own when some cells (which could be thought of as pacemakers) perceive adverse environmental conditions and begin releasing cyclic AMP. This triggers a chain reaction of the other amoebae in the colony to release the chemical and in response they would gravitate toward regions of highest concentrations of cyclic

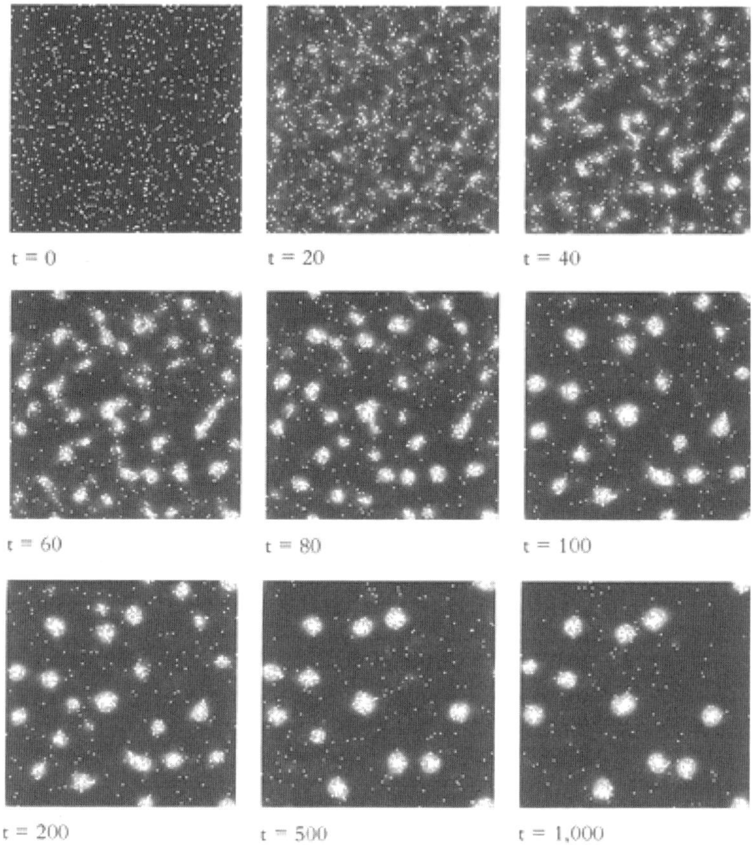

Figure 6.12 The slime mold simulation of Mitchel Resnick starting with 1000 cells at various time steps. It is thought that eventually one aggregate will form.

AMP in a process known as *chemotaxis*. This leads to the formation of small clusters which become larger with the final result being the aggregrate slime mold organism. *Dictyostelium discoideum* illustrates how in Nature a self-organizing system can be formed from the simple interaction of autonomous agents. A comprehensive discussion of various models of slime mold aggregation can be found in the paper by Alan Garfinkel [1987]. A computer simulation that captures some of the salient features of slime mold aggregration has been developed by Mitchel Resnick of MIT and is discussed in his book, *Turtles, Termites, and Traffic Jams* [1994]; see Fig. 6.12.

Figure 6.13 The growth pattern of the bacterial colony *P. dendritiformis*. The size of the colony is approximately 5 cm and the number of bacteria is roughly that of the number of people on Earth.

Rather interestingly, the cyclic AMP released by the amoebae washes through the colony in waves and spirals that can be described by the hodgepodge machine (Section 5.1.3). However, this particular feature does not appear to be essential to the aggregration process, as was demonstrated by Prigogine and Stengers [1984], although Camazine *et al.* [2003] point out that it may serve some adaptive purpose in the case of the slime mold.

6.3.3 Bacterial Colonies

Although the slime mold feeds off bacteria on the forest floor and one does not normally associate much in the way of intelligence with regard to bacteria (although they can get the better of us humans from time to time), researchers are finding a collective "intelligence" even at this level. We have already referred to the sophisticated notion of quorum sensing in bacterial colonies in our discussion of the density problem (Section 3.10). In Fig. 6.13, we see the highly structured self-organized growth pattern of *Paenibacillus dendritiformis*. Other self-organized patterns can be found in the fascinating paper by Eshel Ben-Jacob *et al.* [2004], and stunning color images on the book's website. Quoting from Ben-Jacob *et al.*:

> Here we reason that bacterial chemical conversations also include assignment of contextual meaning to words and sentences (semantic) and conduction of dialogue (pragmatic) — the fundamental aspects of linguistic communication... .
>
> Using these advanced linguistic capabilities, bacteria can lead rich social lives for the group benefit. They can develop collective memory, use and generate common knowledge, develop group identity, recognize the identity of other colonies, learn from experience to improve themselves, and engage in group decision-making, an additional surprising social conduct that amounts to what should most appropriately be dubbed as social intelligence.

This remarkably illustrates that self-organization and complexity have arisen from the very first and simplest living organisms on Earth. It seems that a highly advanced civilization has been living all along right under our noses! And since they have been around a lot longer than we have, perhaps there is even something to be learned from their social interactions.

6.3.4 Langton's Ant

When you have seen one ant, one bird, one tree, you have not seen them all.

— *E. O. Wilson*

In Christopher Langton's 1986 study of artificial life, one of the things he considered was the activities of ants because they can exhibit complex behavior arising from very simple interactions with their environment. To this end he created *virtual ants* or *vants* that play out their lives in a cellular array. These were independently discovered by Greg Turk, although in the more general setting of Turing machines moving on a two-dimensional grid, where they are known as *turmites* (see A. K. Dewdney's *Scientific American* article [1989b]), and also by E. G. D. Cohen [1992] in the setting of lattice gas diffusion models.

To experiment with a variety of ant and turmite creatures, see the website `http://web.comlab.ox.ac.uk/oucl/prospective/ugrad/csatox/demo/turmites.htm`.

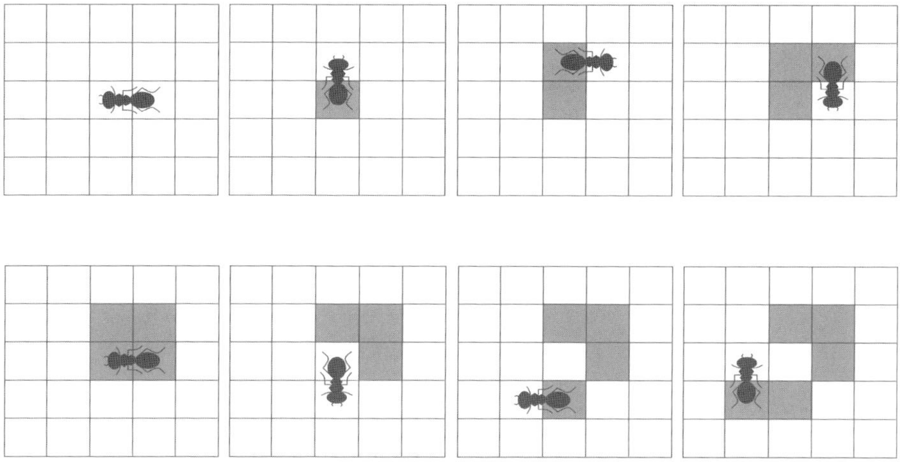

Figure 6.14 The first eight steps taken by Langton's ant.

Let us start with an initial configuration of all white cells (state 0), although a random initial condition of white and black cells (state 1) is also possible. Consider a solitary (v)ant on a central white cell that is facing in an E-W (or N-S) direction that advances one cell forward. It now behaves in the following manner:

- The ant changes any white cell it lands on to black, then turns 90° to the right and moves one cell forward.
- The ant changes any black cell it lands on to white, then turns 90° to the left and moves one cell forward.

Thus a vant operates as a virtual Turing machine on a two-dimensional tape and as such is not a true cellular automaton in that all cells are not updated simultaneously by a local rule.

It can be seen in Fig. 6.14 that the ant quickly starts to run into cells that have already been visited and for the first 500 or so steps creates somewhat symmetrical patterns with 180° rotational symmetry, returning periodically to the central square. It then begins to move in a very random fashion for thousands of steps. The long term evolution of the vant however is quite unexpected and nothing short of wonderous.

Eventually, after more than 10,000 time steps have elapsed, an intricate diagonal "highway" is constructed by the ant that consists of a sequence of 104 steps that then shifts one cell horizontally and vertically (Fig. 6.15) and then repeats this process endlessly.

Figure 6.15 Langton's ant after 11,000 steps with the infinite highway well under construction in the north-east direction. This same structure was independently discovered by Greg Turk and appears in *Scientific American*, September 1989, p. 126.

Although it has thus far not been proven that the ant will always build a diagonal highway, empirical evidence suggests that this is indeed the case even if the initial grid contains a finite number of black squares!

Various generalizations of the vant have been considered. The simplest is to change the rectangular array to some other shape such as a hexagonal grid (see the website of A. Gajardo: `http://www.ing-mat.udec.cl/~anahi/langton/general.html`). See Fig. 6.16.

Another way to generalize the virtual ant is to allow each of the cells of the array to exist in n states instead of just two, say states (or colors) $0, 1, ..., n-1$. If the ant lands on a cell in state k, it changes it to state $k+1 \pmod{n}$, so that each state value is increased by 1 and state $n-1$ changes to state 0. The direction the ant is to turn is governed by a fixed *rulestring* of n symbols consisting of **0**'s and **1**'s. If the ant lands on a cell that is in state k, one then checks the $(k+1)$st symbol of the rulestring, and if it is **1**, the ant turns right 90° and moves one cell

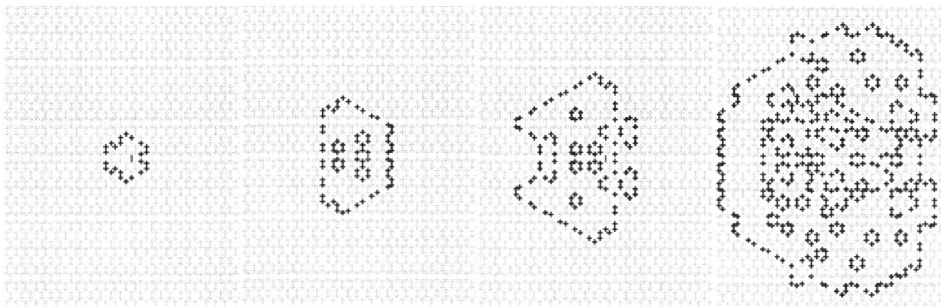

Figure 6.16 The Langton ant on a hexagonal array showing bilateral symmetry in an enlarging pattern. Time steps are at 72, 810, 2600, and 15,600.

forward, and if it is **0**, the ant turns left 90° and moves one cell forward. The rulestring also serves to give an identification number in base 2 so that the ant with rulestring **1001** is called ant 9. Langton's original ant is ant 2 since its rulestring is **10**.

Generalized ants were considered by Greg Turk as remarked above and independently by Leonid Bunimovich and S. E. Troubetzkoy [1994]. Although there are not many definite results that have been proven about generalized ants or even Langton's original ant, for that matter, we do have the following beautiful theorem:

Theorem 3 (Bunimovich-Troubetzkoy Theorem) *The time evolution of the generalized ant is unbounded if the rulestring contains at least one* 0 *and one* 1.

That is to say, the complete trajectory of the ant is not contained within the bounds of any finite array of cells, no matter how large. The argument goes like this. Let us suppose that we are dealing with the Langton ant (rulestring **10**), although the proof (given in Bunimovich-Troubetzkoy [1992]) also applies to the general case. Note first that the cell pattern the ant has created together with its current position and heading uniquely determines the ant's future, and the same is true for its past because the rule governing its behavior is completely reversible. Now if we assume to the contrary that the trajectory of the ant is actually bounded, it follows that at some stage the ant will repeat the same pattern sequence, position and heading because these are only finite in number. By reversibility, the ant's past must be a repetition of this pattern and, clearly, so must the ant's future behavior. In other words, the ant exhibits periodic behavior visiting the same collection of cells infinitely often and nothing more.

Because the ant always turns 90° to the left or right after it enters a cell, if it enters horizontally, it leaves vertically, and if it enters vertically, it leaves horizontally. Thus (depending on the ant's starting orientation), all the cells can be labeled either h (entered horizontally) or v (entered vertically) with contiguous cells taking the opposite designation in a checkerboard fashion. Now there will be one cell which we call M that is as far to the right and top of the array as is possible. Thus any cells to the top or right of M are not visited by the ant. If M happens to be an h cell, it must have been visited horizontally from the left. It must turn 90° to the right, implying that M must have been a white cell which the ant turns to black. On the next cycle of the ant, since M is now black and entered horizontally from the left, the ant must turn 90° to the left and exit vertically out the top of M. But no cells above M are visited, resulting in a contradiction. If M is a v cell, a similar contradiction arises. Thus it must be the case that the trajectory of the ant is unbounded, proving the theorem.

Some trajectories of the generalized ant are trivial, such as the one with rulestring **11**, which simply goes around and around a 2×2 square. Other trivial rulestrings are **111**, **1111**, ... Rulestring **01** is just the mirror image of the Langton ant. For ants with rulestring of length $n > 2$, take cell state 0 to be white and cell state $n - 1$ to be black, using shades of gray for the intermediate cell states with all cells initially in state 0. In what follows are some results of computer simulations by James Propp [1994].

Ant 4 (**100**) creates various symmetrical patterns with bilateral symmetry (Fig. 6.17) but then resorts to random behavior without any evidence of settling down to some regular structure as was the case with ant 2.

Ant 5 (**101**) is similar to ant 2 but does not seem to create a highway. However, ant 6 (**110**) does create a highway after 150 steps, but this time the structure is composed of a sequence consisting of only 18 steps, or "period" 18 (Fig. 6.18), unlike the period 104 highway of the Langton ant 2.

Ant 8 (**1000**) does not build highways nor do any other regular patterns emerge. Ants 9 (**1001**) and 12 (**1100**) create ever larger and more elaborate patterns that very often are bilaterally symmetric the moment

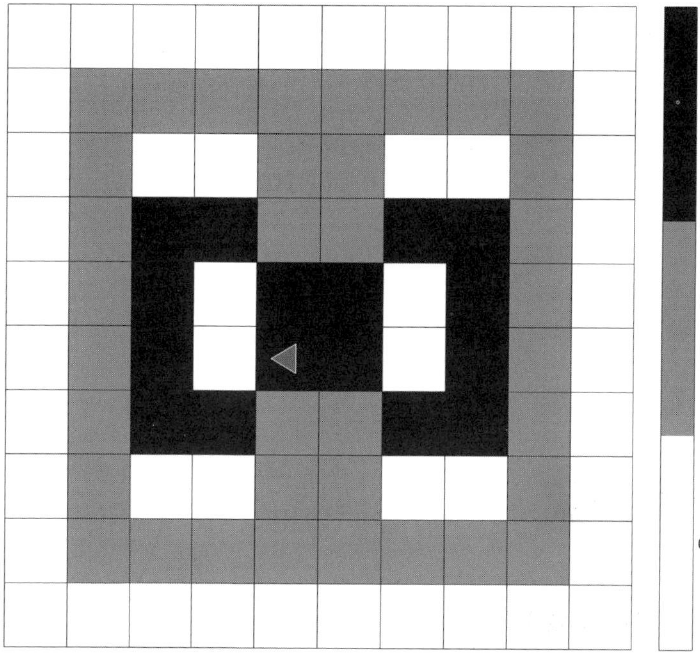

Figure 6.17 The bilateral symmetry exhibited by ant 4 at step 236 of its evolution.

the ant returns to the initial starting cell, a phenomenon already noted by Greg Turk. Indeed, Bernd Rummler has proved that ant 12 will continue to create bilaterally symmetric structures, and this turns out to be the case for some other ants as well (see Gale *et al.* [1995]). Yale university students and alumni wll be pleased to know that ant 9 creates a respectable image of a bulldog after 38,836 steps (Fig. 6.19).

Ant 10 (**1010**) has the same behavior as ant 2 (**10**) only with a four-color trail instead of one with two colors. This is due to the general principle that, if a rulestring is built up of repetitions of a shorter rulestring, then the behavior of the former will be that of the latter.

Ant 13 (**1101**) wanders around in a random fashion for about 250,000 steps but then builds a highway having period 388. Ant 14 (**1110**) is rather interesting in that it possesses some features of ants 2 and 6. It creates a highway, but the period is 52, that is, half that of the Langton highway, and the appearance is like that of ant 6. Experimental evidence of Olivier Beuret and Marco Tomassini [1997] suggests that whenever

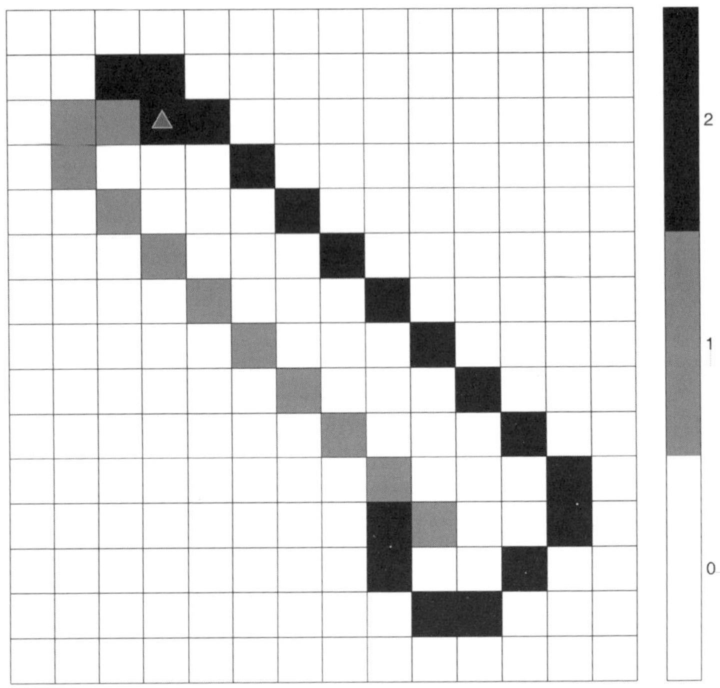

Figure 6.18 The highway exhibited by ant 6 consisting of a sequence of black cells of period 18.

the triple **110** is preceded by **1**'s, as in **1110**, **11110**, ..., that particular ant will also build a highway.

Employing 5-bit rulestrings, unlike ants 9 and 12, there were no ants that created bilateral symmetric structures. However, Propp found one new feature in ant 27 (**11011**), namely, an increasing spiral pattern. With rulestrings of length 6, Propp found that ants 33, 39, 48, 51, 57, and 60 each exhibited bilateral symmetry, all of whose numbers are divisible by 3. The reason why this is so can be found in Gale *et al.* [1995].

One truly remarkable aspect of all of this is the fact that, even though we know the explicit rule by which an ant's behavior is governed, the emergence of a highway after 10,000 steps for ant 2 or 250,000 steps for ant 13 is completely unexpected. Nothing in their prior behavior gives any hint that the long term future of these ants lies in periodic highway building, nor is there any hint in the simplistic rulestring. Moreover, it

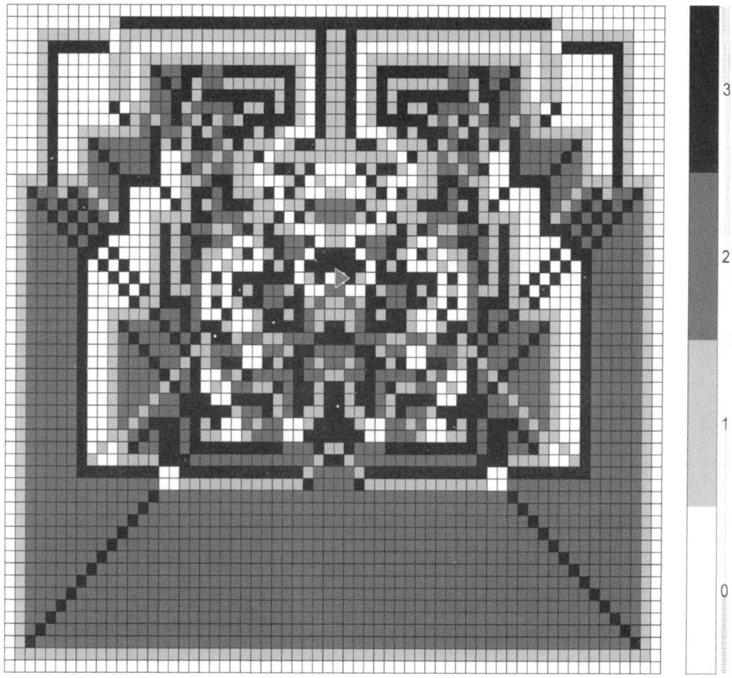

Figure 6.19 The bilaterally symmetric pattern exhibited by ant 9 at step 38,836 of its evolution.

is quite possible that the question of whether certain ants, such as ant 4, say, ever produce a highway is undecidable and hence their long term behavior can only be ascertained by living it (see Stewart [1994]).

6.3.5 Multi-Ant Systems

Empirical investigations of systems containing two three and four ants have been considered by graduate students Malcolm Walsh and Dylan Hirsch-Shell. The reader is invited to experiment with multi-ant systems and send interesting phenomena to the website (see Preface) for this book. Details of each ant's initial position and facing direction can also be found on the website.

Interactions between two ants can have varying consequences. In dealing with multi-ant systems, there is always the possibility that at any particular time step two (or more) ants will meet up at the same square. In this instance, a right-of-way rule must be agreed upon and by numbering the ants $1, 2, \ldots,$ we agree that 1 precedes 2, 2 precedes

Figure 6.20 The production of individual orthogonal Langton highways by two ants showing their evolution after 6750 steps. Grid lines have been suppressed.

3, and so forth. This numbering, which we will write as ant1, ant2, ..., should not be confused with that in the preceding section. In our first experiment with two ants (Fig. 6.20) it was found that highway production is initiated much earlier, although not simultaneously, than with the original single ant, which took more than 10,000 steps to produce a Langton highway. Parallel highways have also been observed.

Of course, with two ants there is always the possibility that they will undo each other's work and return to their initial positions. This phenomenon indeed ocurrs and was found to be not so uncommon. This results in periodic behavior (Fig. 6.21).

The next sequence (Fig. 6.22) depicts an extraordinary two-ant interaction. Ant1 (bottom initial position) begins building a traditional Langton highway on its own (second image). In the third image, ant2 has sprinted up the highway and caught ant1, joining it in close proximity near the top. Having done this, ant2 returns to the base of the

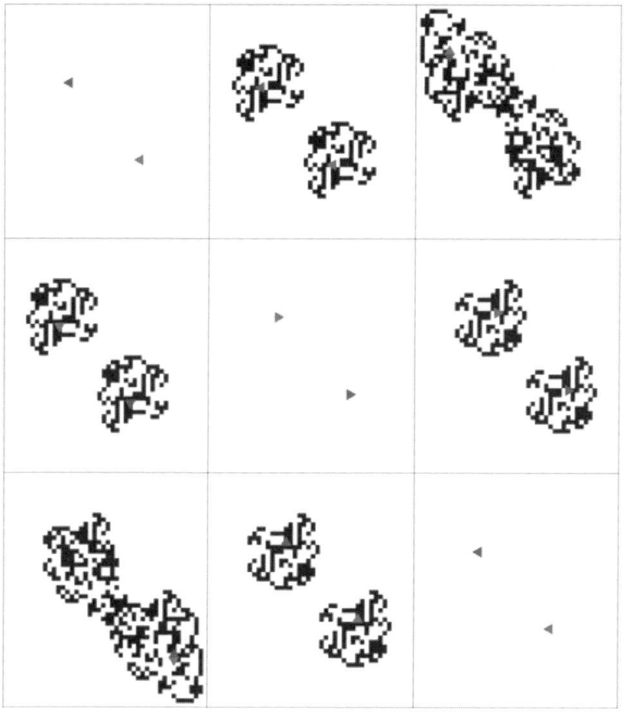

Figure 6.21 The interaction of two ants leading to the mutual destruction of all their work and their eventual return to the initial positions and orientations, which results in 2-cycle behavior. In the central square, the ants are not only facing 180° opposite to their initial orientation but have swapped positions as well. The time evolution proceeds left to right on each of the three rows.

figure to roam around some more, while ant1 carries on with some messy construction (fourth frame). After some further construction activities of ant1, it descends to the base to rejoin ant2 (fifth image, 4904 steps). Now comes the truly remarkable part. At the very next step, things start to reverse themselves, with each ant, respectively, undoing the work of the other ant, both top and bottom (sixth image). After 7969 steps (seventh image), we find ourselves back in a similar situation as the third image, only this time ant1 has joined up with ant2 near the top of the highway. Finally, ant1 descends the highway to undo the work of ant2 at the base and ant2 undoes the highway initiated by ant1 (image 8), with the two ants returning home with reversed starting positions

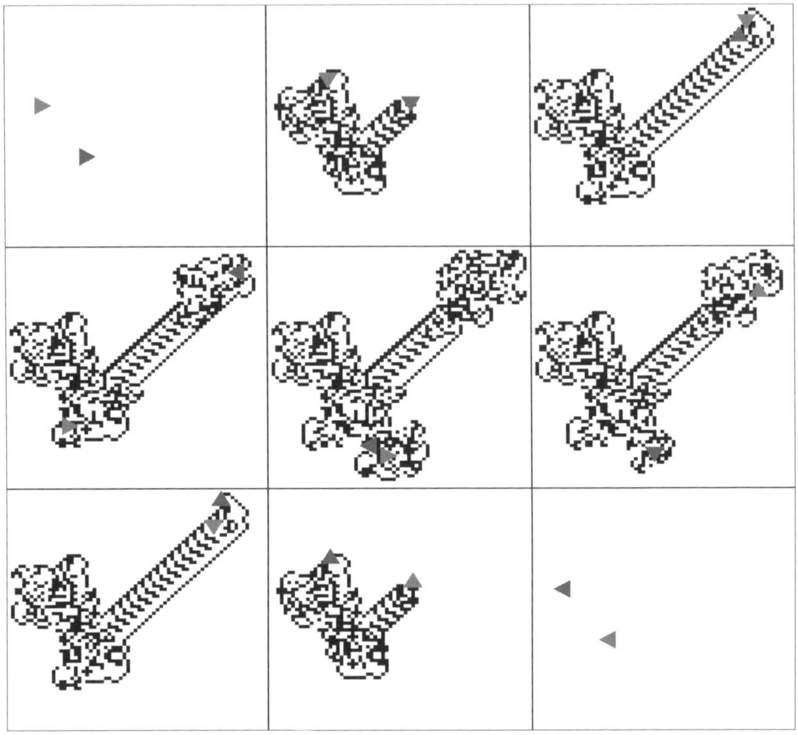

Figure 6.22 A sequence depicting a two-ant interaction with remarkable construction and destruction aspects. See text for explanation.

and orientations rotated by 180°. This is analogous to the first cycle of the two ants in the preceding example.

On the surface, this certainly appears to be some sort of "intelligent" behavior, but we know that the ants are both just following the simple, dare we say "dumb," rule set down by Langton.

In the preceding, the ants' orientations at the final stage are rotated by 180°. Continuing to let the ants run on, they interact once more but ultimately again undo each other's work, this time returning to their original positions and orientations. This results in a periodic 2-cycle of frenzied activity. Significantly, this experiment and the preceding one demonstrate that: *A system of two ants can result in their behavior being entirely bounded,* in contrast to the necessarily unbounded behavior of one ant established in the Bunimovich-Troubetzkoy theorem. After running these experiments and discovering this bounded behavior we found that this phenomenon has also been noted in the book by Chopard

Figure 6.23 The new type of cooperative highway found by Malcolm Walsh, in this instance heading in a NNW direction, made by two ants having period 102. Note the different structure to the single-ant Langton highway.

and Droz [1998]. Of course, as we have seen above, there are two-ant systems that exhibit unbounded behavior as well. Similarly, four-ant and eight-ant systems have been found that exhibit analogous 2-cycle bounded behavior for all time.

In a further suitably chosen two-ant scenario, the ants begin to make a *cooperative* highway after around 500 steps which seems to be a heretofore unobserved phenomenon (Fig. 6.23). This cooperative highway has a somewhat different structure than the normal Langton highway and has a period of 102 compared to the single-ant Langton highway which has a period of 104. So, these "mindless" ants, just like humans, can not only destroy one another's work but also cooperate on construction projects as well.

Another interaction that can occur is that one ant building a highway will meet the highway of another ant. One scenario is that the ants cease their highway construction and revert to more disorderly behavior. On the other hand, it is possible for the ant hitting the highway of the other to actually proceed out the other side of the first ant's highway, as in Fig. 6.24.

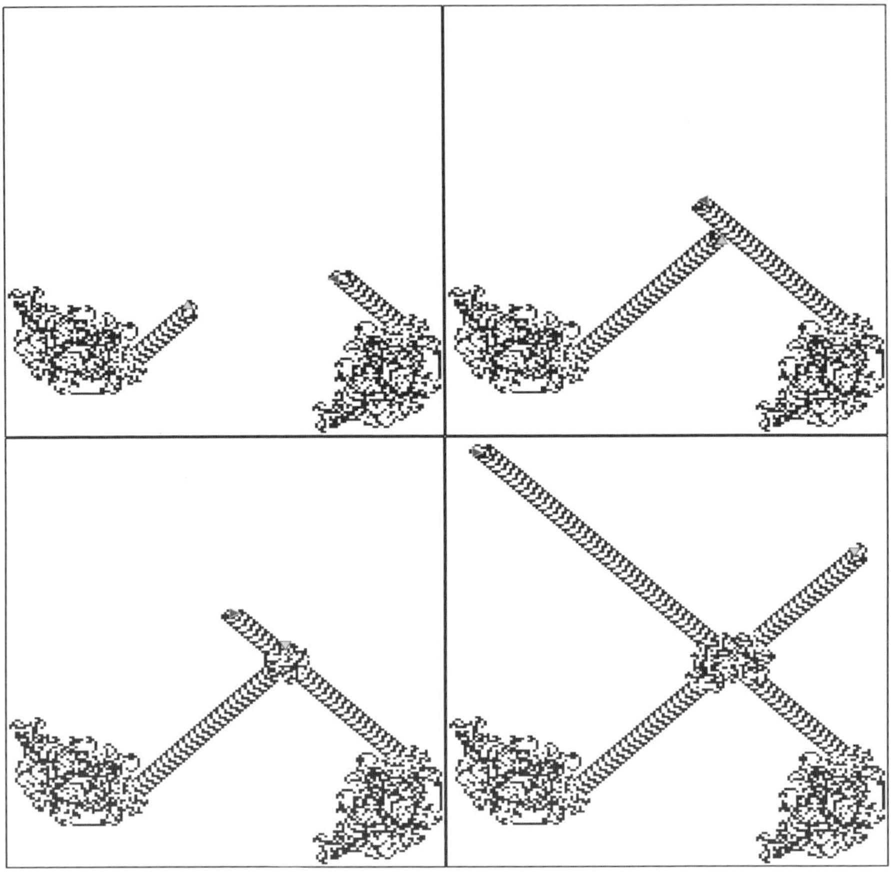

Figure 6.24 In a two-ant system we can have the intersection of two highways and the continued construction of both, with each coming out the other side of the other.

In three-ant systems, experiments have shown that multiple highway systems can be built and unbuilt; however, it seems to be the case that these systems are again unbounded as for a single ant. Further empirical evidence on the matter needs to be gathered. Once again, the 102-step cooperative highway structure can be created by two of the ants with suitable initial conditions. After around 3000 steps ant 3 will begin creating its own Langton highway in nearly the opposite direction, never to interact with the other two ants again (Fig. 6.25). The construction of three separate Langton highways has also been observed. Will n-ant systems, where n is odd, always be unbounded?

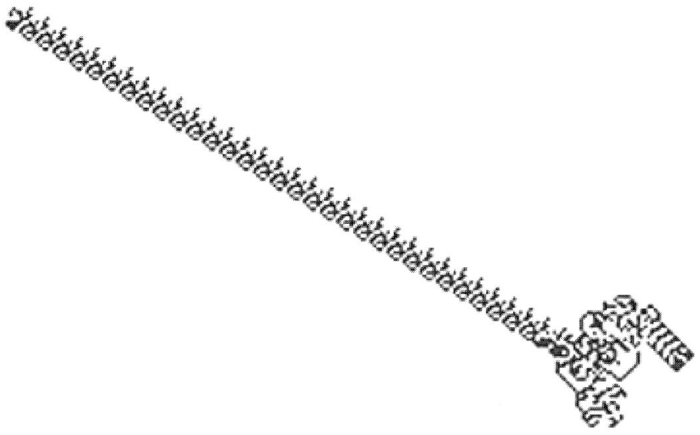

Figure 6.25 The three-ant system after 3250 steps that shows the construction of the period 102 cooperative highway by two of the ants (left) and the third ant starting to make its own Langton-highway (right).

Four-ant systems display similar characteristics to two-ant systems with most leading to four unbounded highways, and as mentioned above, along with eight-ant systems, some have been shown to remain bounded. This occurs if the ants are symmetrically placed so that they will pair off with the result that the highway constructed by one of the pair will eventually be undone by the other.

6.3.6 Traveling Salesman Problem

Virtual ants have found a very real application in finding a solution to the famous *traveling salesman problem*. An early instance of it was presented at a mathematical colloquium in Vienna by the mathematician Karl Menger in 1930 as *Das botenproblem* ("messenger problem").

A number of cities are spread out across a two-dimensional map and a traveling salesman is required to find the shortest route that visits each of the cities once and returns to the original site. Of course, the salesman knows the distance between each pair of cities. The difficulty arises from the fact that for n given cities there are $n!$ possible routes. However, we need only consider half this number as every route can be traversed in reverse order, which is essentially the same route, and since we can start in any given city, the number of different routes is further reduced by a factor of n. This still leaves $(n-1)!/2$ different

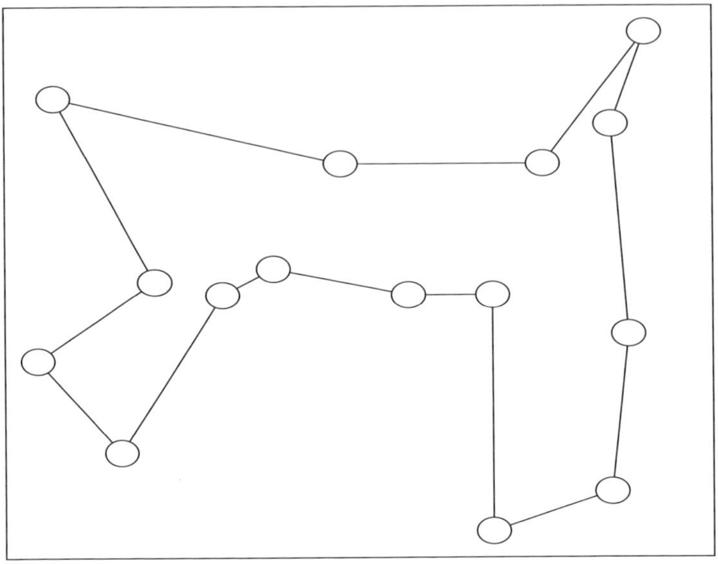

Figure 6.26 A 15-city optimal route solution to the traveling salesman problem found by brute force calculation.

routes to choose from. So, for example, with only 15 cities there are over 43 billion different possible routes (see Fig. 6.26). There is simply no efficient mathematical algorithm that can find the shortest route in all cases, and this is the sort of problem that computer scientists call *NP complete*. (Sadly for physicists, Sorin Istrail in 2000 showed that another such problem is finding a mathematical solution for the three-dimensional case of the Ising spin model.) The traveling salesman problem has many practical applications, involving traveling not only between cities but to telecommunication networks and even to the machine punching of holes into circuit boards.

There are various schemes to find optimal routes, or at least near optimal ones, and one recent one due to Marco Dorigo and Luca Gambardella [1997] involves virtual ants. Their work was inspired by the behavior of colonies of *real* ants.

In a stunning demonstration of self-organization, real ants are capable of finding the shortest route from their nest to a food source without using any visual cues from their environment, and certainly without any realization of what they are doing. Even if an obstacle is put in their path, the ants will soon find the next shortest path to the food source.

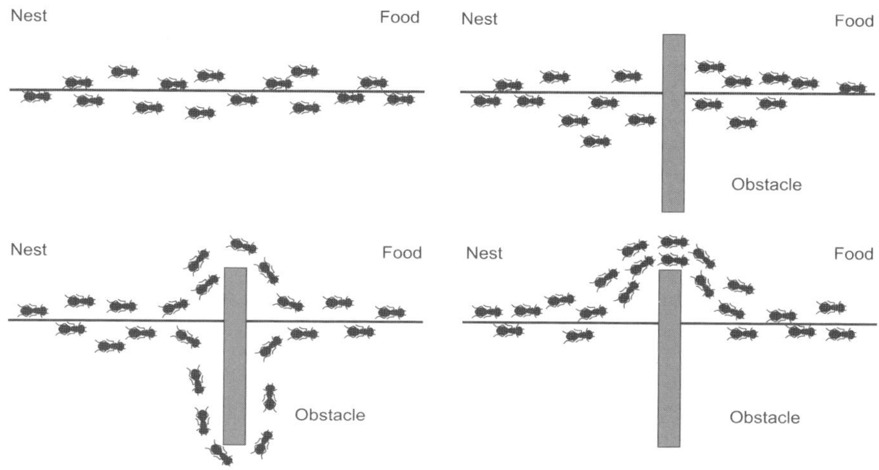

Figure 6.27 Ants following a pheromone trail will soon discover the shortest route to a food supply even if an obstacle is put in their way as the shorter route will accumulate more pheromone.

They manage to do this because each ant lays a chemical pheromone trail whenever it is carrying food. Coupled to this is the ants' preference to follow a trail that is richer in pheromones over one that is not (see Ermentrout and Edelstein-Keshet [1993] for a simple model of this behavior). Since an ant who has taken a shorter route to a food source will return with it somewhat sooner than an ant who travels to a more distant source, the former ant's trail becomes reinforced by other ants and accumulates more pheromone per unit of time than longer routes. Soon the latter die out, due to evaporation, and all ants will follow the shortest one (Fig. 6.27).

Dorigo and Gambardella realized that the ants finding the shortest route was a phenomenon that could be put to use in finding optimal routes in the traveling salesman problem. In the electronic version, the cities are connected by pathways and an army of virtual ants is allowed to travel along them leaving a virtual "pheromone" trail as they go from city to city. The ants are given the propensity to prefer trails with a lot of pheromone along pathways that connect nearby cities. Once all the ants have completed their tour of duty, the ant with the shortest route has its route further enhanced with pheromone so that, the shorter the route, the more enhancement it receives. This process is then repeated many times even allowing for the evaporation of pheromone at a steady rate as happens in real ant colonies (although the rate adopted by the researchers was much higher). The resulting *ant colony system* (*ACS*) can achieve near-optimal to optimal solutions to the traveling salesman problem and is as good or better than other competing systems having already been applied to telecommunications networks. Ant colony behavior is providing impetus to other fields such as robotics, data sorting, truck routing, and office design.

Artificial ants have had another remarkable triumph. In experiments by Moyson and Manderick [1988], nests of ants were modeled that had more than one food source. The artificial ants once again found minimal paths to these sources but also found minimal paths *between food sources.* This was originally thought to be the result of a "bug" in the system, but when checked with nests of real ants, this same behavior was also observed.

Thus there is a lot to be learned from the humble ant, both virtual and real!

APPENDIX A

Wolfram's 256 elementary cellular automata, all starting with one black cell.

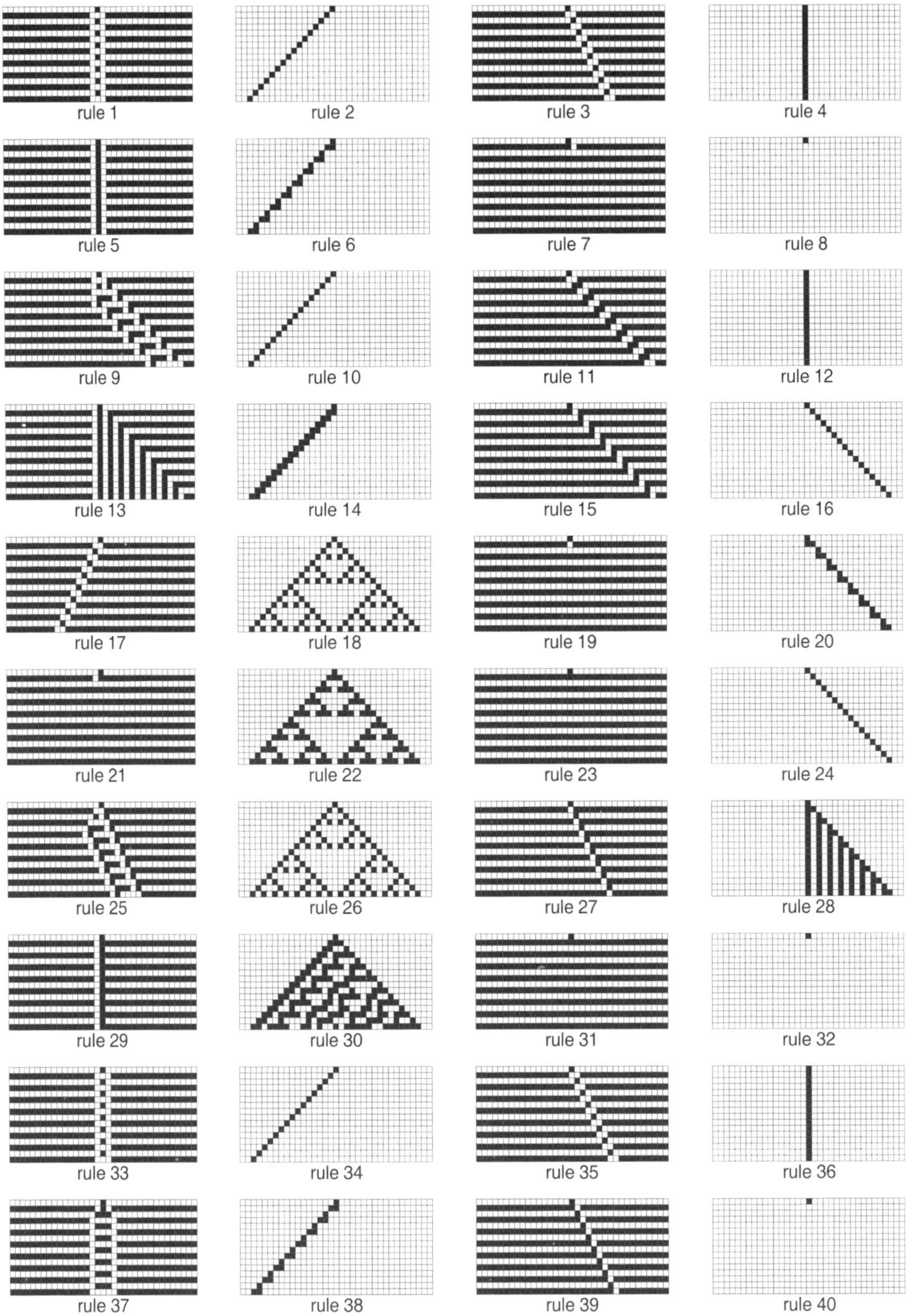
rule 1
rule 2
rule 3
rule 4
rule 5
rule 6
rule 7
rule 8
rule 9
rule 10
rule 11
rule 12
rule 13
rule 14
rule 15
rule 16
rule 17
rule 18
rule 19
rule 20
rule 21
rule 22
rule 23
rule 24
rule 25
rule 26
rule 27
rule 28
rule 29
rule 30
rule 31
rule 32
rule 33
rule 34
rule 35
rule 36
rule 37
rule 38
rule 39
rule 40

Figure A.1

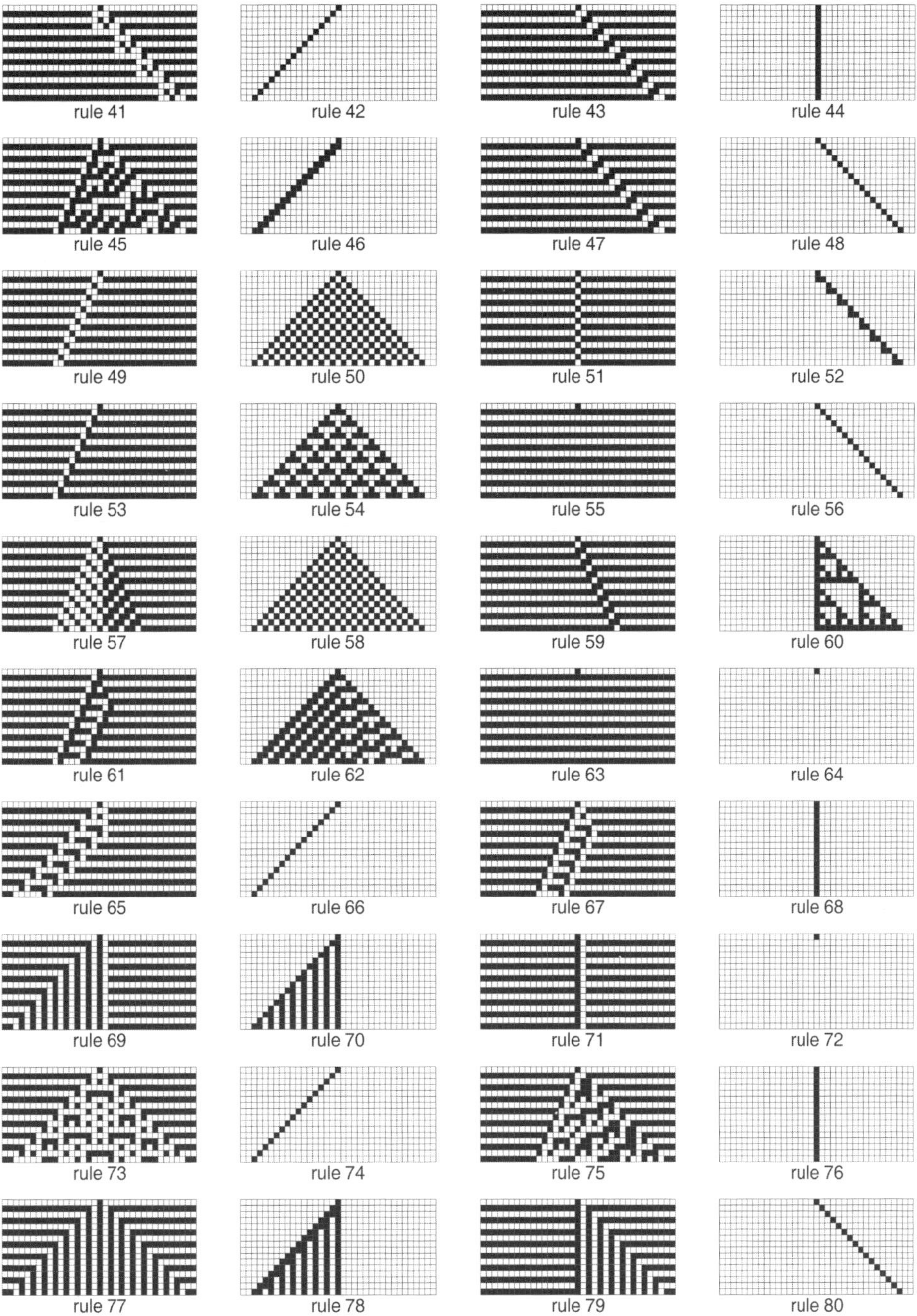
rule 41
rule 42
rule 43
rule 44
rule 45
rule 46
rule 47
rule 48
rule 49
rule 50
rule 51
rule 52
rule 53
rule 54
rule 55
rule 56
rule 57
rule 58
rule 59
rule 60
rule 61
rule 62
rule 63
rule 64
rule 65
rule 66
rule 67
rule 68
rule 69
rule 70
rule 71
rule 72
rule 73
rule 74
rule 75
rule 76
rule 77
rule 78
rule 79
rule 80

Figure A.2

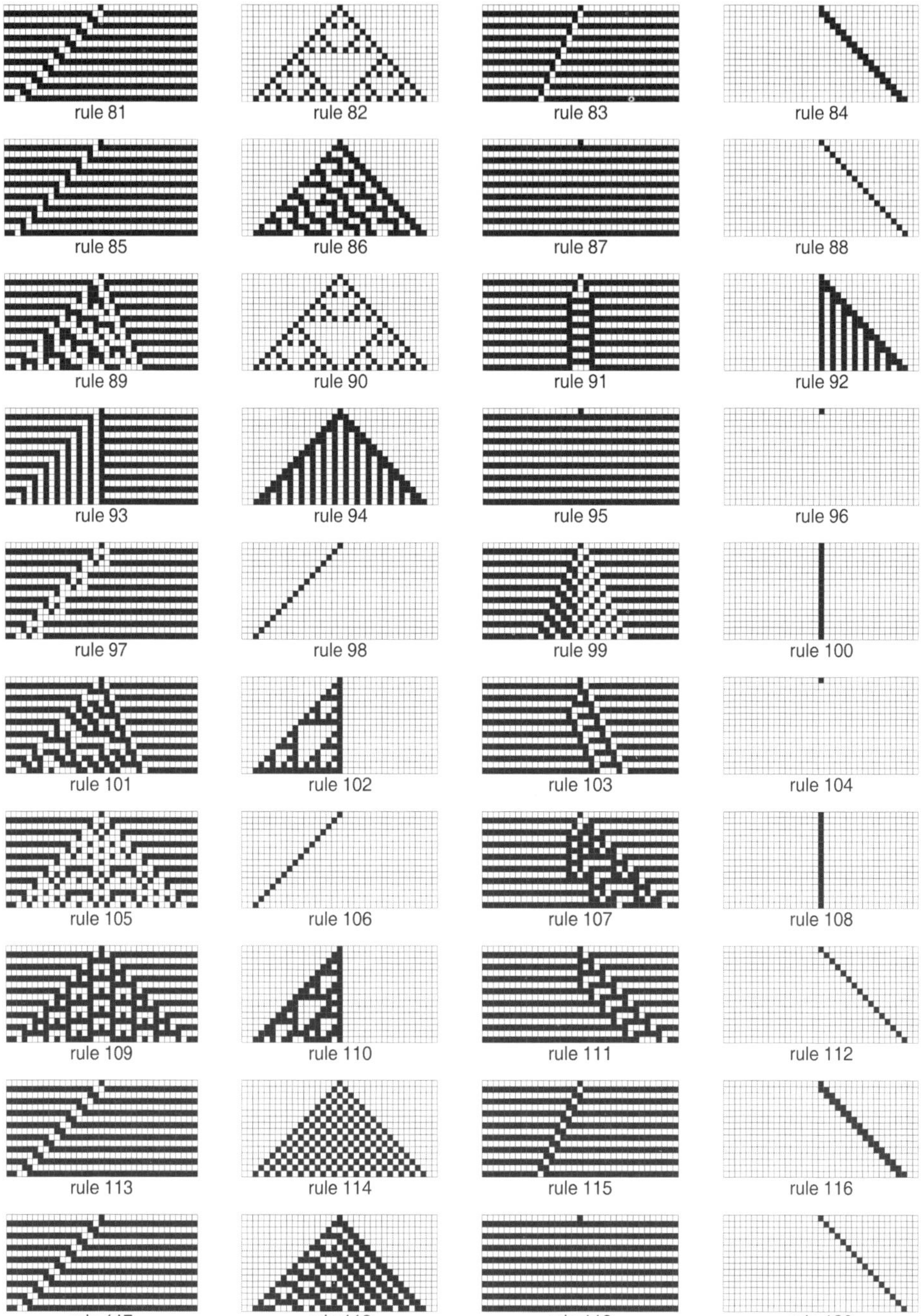
rule 81
rule 82
rule 83
rule 84
rule 85
rule 86
rule 87
rule 88
rule 89
rule 90
rule 91
rule 92
rule 93
rule 94
rule 95
rule 96
rule 97
rule 98
rule 99
rule 100
rule 101
rule 102
rule 103
rule 104
rule 105
rule 106
rule 107
rule 108
rule 109
rule 110
rule 111
rule 112
rule 113
rule 114
rule 115
rule 116
rule 117
rule 118
rule 119
rule 120

Figure A.3

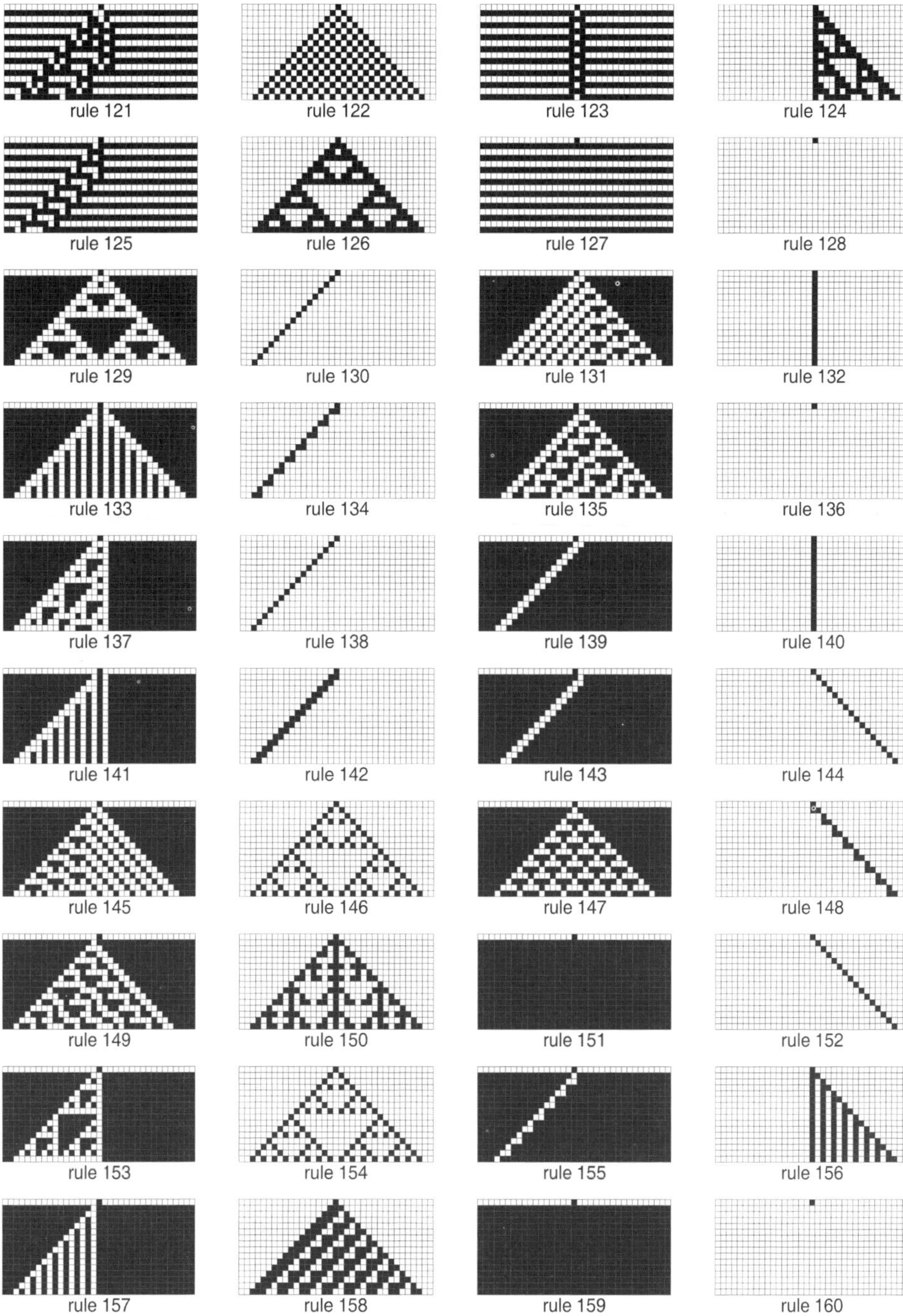
rule 121
rule 122
rule 123
rule 124
rule 125
rule 126
rule 127
rule 128
rule 129
rule 130
rule 131
rule 132
rule 133
rule 134
rule 135
rule 136
rule 137
rule 138
rule 139
rule 140
rule 141
rule 142
rule 143
rule 144
rule 145
rule 146
rule 147
rule 148
rule 149
rule 150
rule 151
rule 152
rule 153
rule 154
rule 155
rule 156
rule 157
rule 158
rule 159
rule 160

Figure A.4

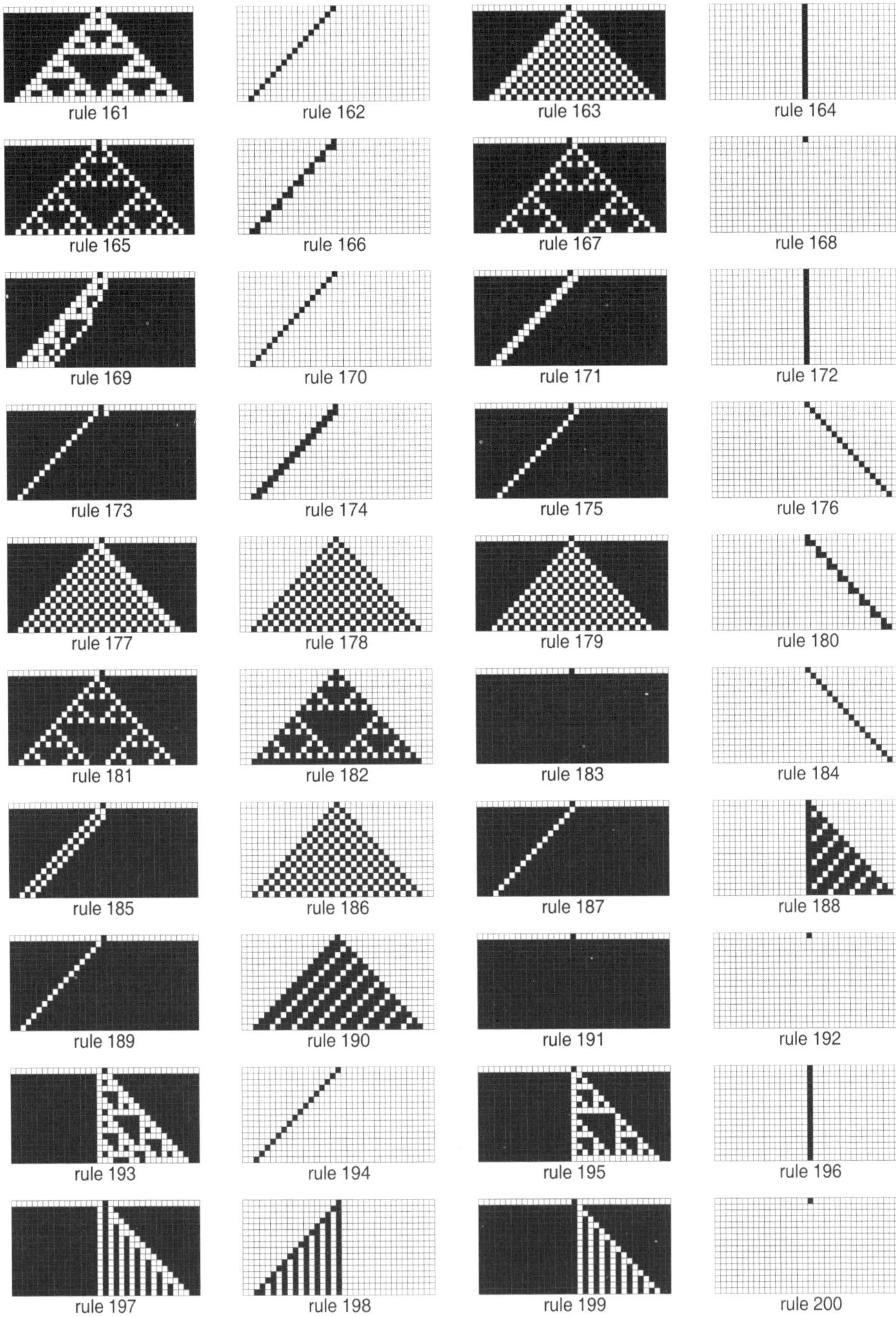
rule 161
rule 162
rule 163
rule 164
rule 165
rule 166
rule 167
rule 168
rule 169
rule 170
rule 171
rule 172
rule 173
rule 174
rule 175
rule 176
rule 177
rule 178
rule 179
rule 180
rule 181
rule 182
rule 183
rule 184
rule 185
rule 186
rule 187
rule 188
rule 189
rule 190
rule 191
rule 192
rule 193
rule 194
rule 195
rule 196
rule 197
rule 198
rule 199
rule 200

Figure A.5

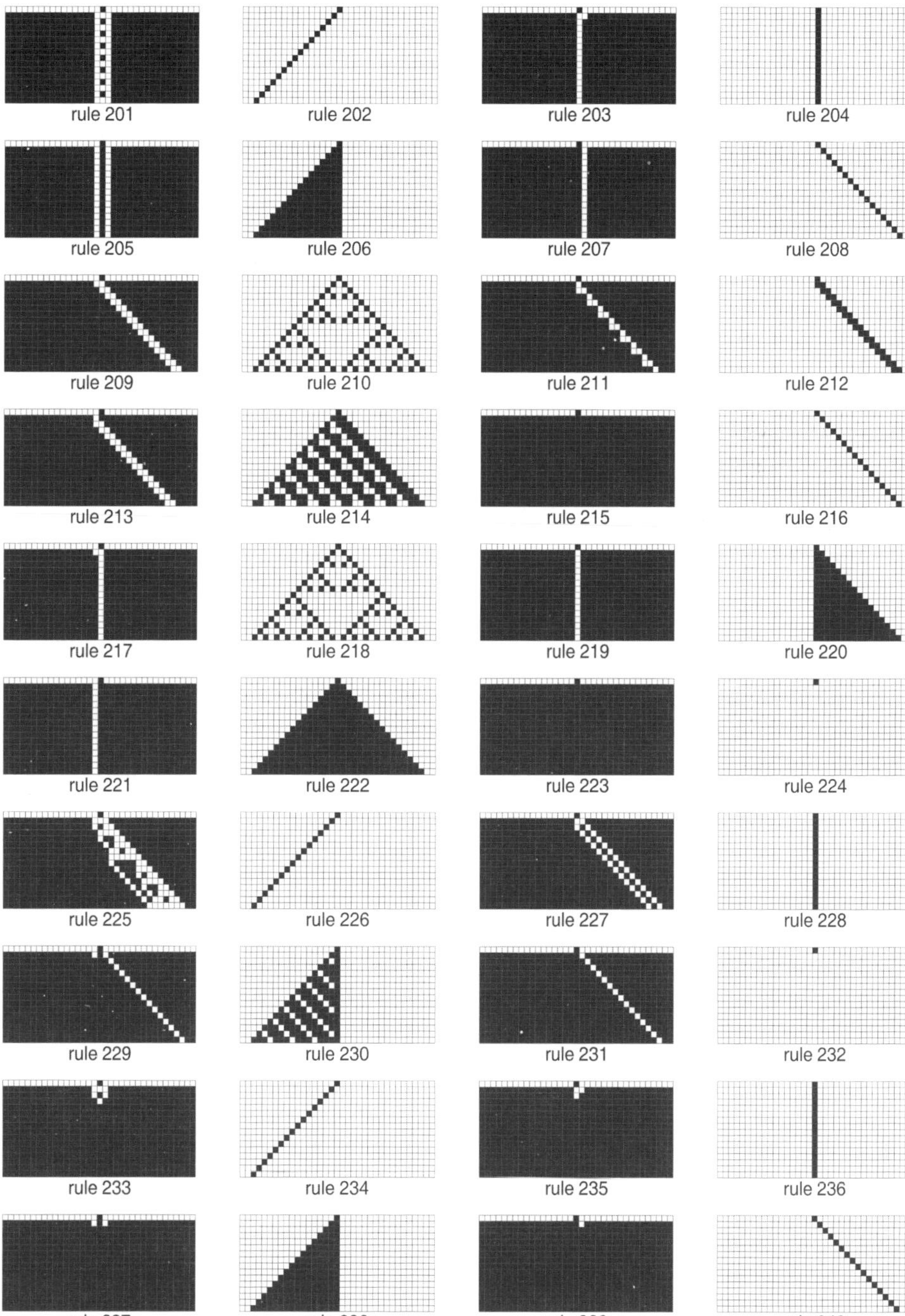
rule 201
rule 202
rule 203
rule 204
rule 205
rule 206
rule 207
rule 208
rule 209
rule 210
rule 211
rule 212
rule 213
rule 214
rule 215
rule 216
rule 217
rule 218
rule 219
rule 220
rule 221
rule 222
rule 223
rule 224
rule 225
rule 226
rule 227
rule 228
rule 229
rule 230
rule 231
rule 232
rule 233
rule 234
rule 235
rule 236
rule 237
rule 238
rule 239
rule 240

Figure A.6

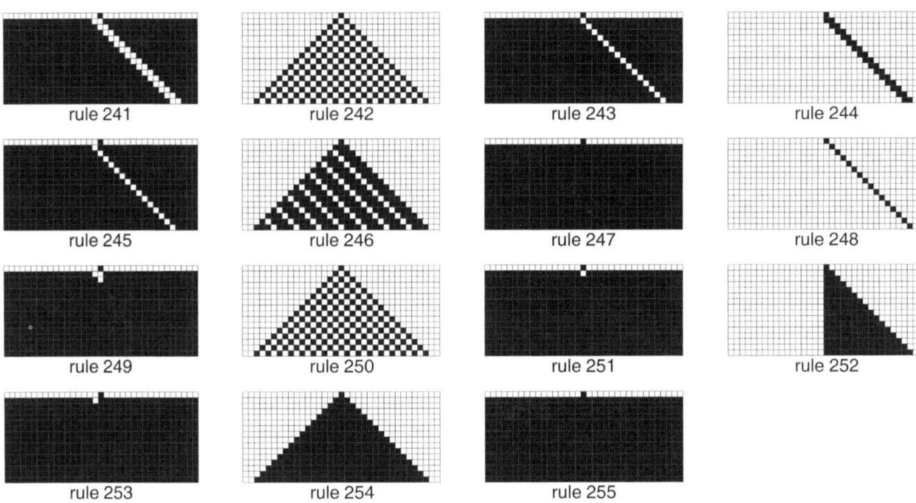
rule 241
rule 242
rule 243
rule 244
rule 245
rule 246
rule 247
rule 248
rule 249
rule 250
rule 251
rule 252
rule 253
rule 254
rule 255

Figure A.7

REFERENCES

Adami, C., *Introduction to Artificial Life*, Springer-Verlag, New York, 1998.

Amoroso, S. and Patt, Y. N., Decision procedures for surjectivity and injectivity of parallel maps for tessellation structures, *J. Comp. Syst. Sci.* **6** (1972), 448–464.

Auslander, D., Guckenheimer, J., and Oster, G., Random evolutionary stable strategies, *Theor. Pop. Biol.* **13** (1978), 276–293.

Axelrod, R., *The Evolution of Cooperation*, Basic Books, New York, 1984.

Axelrod, R., The evolution of strategies in the iterated prisoner's dilemma, in *Genetic Algorithms and Simulated Annealing*, L. D. Davis, ed., Morgan Kaufmann, Los Altos, CA, 1987, 32–41.

Axelrod, R. and Hamilton, W. D., The evolution of cooperation, *Science* **211** (1981), 1390–1396.

Bak, P., Tang, C., and Wiesenfeld, K., Self-organized criticality, *Phys. Rev. A* **38** (1988), 364–374.

Balzer, R., An 8-state minimal time solution to the firing squad synchronization problem, *Inform. Control* **10** (1967), 22–42.

Ben-Jacob, E., Becker, I., Shapira, Y., and Levine, H., Bacterial linguistic communication and social intelligence, *Trends Microbiol.* **12** (2004), 267–272.

Bennett, C. H., Brassard, C., and Ekert, A., Quantum cryptography, *Sci. Am.* **269** (1992), 26–33.

Bentley, W. A., *Snow Crystals*, Dover, New York, 1962.

Berlekamp, F. R., Conway, J. H., and Guy, R. K., *Winning Ways for Your Mathematics Plays*, Vol. 2, Academic Press, New York, 1982.

Bersini, H. and Detour, V., Asynchrony induces stability in cellular automata based models, in *Artificial Life IV*, R. A. Brooks and P. Maes, eds., MIT Press, Cambridge, MA, 1994, 382–387.

Beuret, O., and Tomassini, M., Behaviour of multiple generalized Langton's ant, in *Artificial Life V: Proceedings of the Fifth International Workshop on the Simulation and Synthesis of Living Systems*, C. G. Langton and K. Shimohara, eds., MIT Press, Cambridge, MA, 1997, 45–50.

Bilke, S. and Sjunnesson, F., Stability of the Kauffman model, *Phys. Rev. E* **65** (2001), 016129, 5 pp.

Bonabeau, E., Dorigo, M., and Theraulaz, G., *Swarm Intelligence: From Natural to Artificial Systems*, Santa Fe Institute Studies in the Science of Complexity, Oxford University Press, 1999.

Box, G. E. P., Evolutionary operation: A method for increasing industrial productivity, *Appl. Stat.* **6** (1957), 81–101.

Bricmont, J., Science of chaos or chaos of science, in *Flight from Science and Reason*, P. R. Gross, N. Levitt, and M. W. Lewis, eds., John Hopkins University Press, 1997.

Buckingham, D. J., Some facts of life, *Byte* **3** (1978), 54.

Bunimovich, L. and Troubetzkoy, S., Recurrence properties of Lorentz lattice gas cellular automata, *J. Stat. Phys.* **67** (1992), 289–302.

Bunimovich, L. and Troubetzkoy, S., Rotators and periodicity in cyclic cellular automata, *J. Stat. Phys.* **74**, (1994), 1–14.

Butler, S., *Erewhon*, Penguin Classics, 1970.

Camazine, S., Deneubourg, J-L., Franks, N., Sneyd, J., Theraulaz, G., and Bonabeau, E., *Self-Organization in Biological Systems*, Princeton University Press, 2003.

Capcarrere, M. S., Sipper, M., and Tomassini, M., Two-state, $r = 1$ cellular automaton that classifies density, *Phys. Rev. Lett.* **77** (1996), 4969–4971.

Celada, F. and Seiden, P. E., A computer model of cellular interactions in the immune system, *Immunol. Today* **13** (1992), 56–62.

Chaudhuri, P. P., Chowdhury, D. R., Nandi, S., and Chattopadhyay, S., *Additive Cellular Automata: Theory and Applications, Vol. 1*, Wiley–IEEE Computer Science Press, 1997.

Chua, L. O. and Yang, L., Cellular neural networks: Theory, *IEEE Trans. Circuits Systems* **35** (1988a), 1257–1272.

Chua, L. O. and Yang, L., Cellular neural networks: Applications, *IEEE Trans. Circuits Systems* **35** (1988b), 1273–1290.

Chopard, B. and Droz, M., *Cellular Automata Modelling of Physical Systems*, Cambridge University Press, 1998.

Codd, E. F., *Cellular Automata*, Academic Press, New York, 1968.

Cohen, E. G. D., New types of diffusion in lattice gas cellular automata, in *Microscopic Simulations of Complex Hydrodynamic Phenomena*, M. Mareschal and B. L. Holian, eds., Plenum Press, New York, 1993, 137–152.

Cook, M., Universality in elementary cellular automata, *Complex Systems* **15** (2004), 1–40.

Creutz, M., Deterministic Ising dynamics, *Ann. Phys.* **167** (1986), 62–72.

Das, R., Crutchfield, J. P., Mitchell, M., and Hanson, J. E., Evolving globally synchronized cellular automata, in *Proceedings of the Sixth International Conference on Genetic Algorithms*, L. J. Eshelman, ed., Morgan Kaufmann, 1995, 336–343.

Dawkins, R., The evolution of evolvability, in *Artificial Life I,* C. G. Langton, ed., Addison-Wesley, 1989, 201–220.

Dawkins, R., *A Devil's Chaplain*, Phoenix, 2004.

De Boer, R. J. and Hogeweg, P., Growth and recruitment in the immune network, in *Theoretical and Experimental Insights into Immunology*, A. F. Perelson and G. Weisbuch, eds., Vol. 66, Springer-Verlag, New York, 1992.

Denbigh, K. G. and Denbigh, J. S., *Entropy in Relation to Incomplete Knowledge*, Cambridge University Press, 1985.

Deutch, A. and Dormann, S., *Cellular Automaton Modeling of Biological Pattern Formation*, Birkhäuser, Boston, 2005.

Devaney, R. L., *Introduction to Chaotic Dynamical Systems*, 2nd ed., Addison-Wesley, 1989.

Dewdney, A. K., Sharks and fish wage an ecological war on the toroidal planet Wa-Tor, *Sci. Am.*, December 1984, 14–22.

Dewdney, A. K., *The Armchair Universe*, W. H. Freeman, New York, 1988.

Dewdney, A. K., A cellular universe of debris, droplets, defects and demons, *Sci. Am.* August 1989a, 88–91.

Dewdney, A. K., Two-dimensional Turing machines and tur-mites make tracks on a plane, *Sci. Am.*, September 1989b, 124–127.

Dorigo, M. and Gambardella, L. M., Ant colonies for the traveling salesman problem, *Biosystems* **43** (1997), 73–81.

Elaydi, S. N., *Discrete Chaos*, Chapman and Hall/CRC, 2000.

Ermentrout, G. B., Campbell, J., and Oster, A., A model for shell patterns based on neural activity, *The Veliger* **28** (1986), 369–388.

Ermentrout, G. B. and Edelstein-Keshet, L., Cellular automata approaches to biological modeling, *J. Theor. Biol.* **160** (1993), 97–133.

Feynman, R., *The Character of Physical Law*, Messenger Lecturers – 1964, MIT Press, Cambridge, MA, 1967.

Flake, G. W., *The Computational Beauty of Nature*, MIT Press, Cambridge, MA, 1998.

Flood, M. M., Some experimental games, Technical Report RM-789-1, The Rand Corporation, Santa Monica, CA, 1952.

Franklin, S. and Graesser, A., Is it an agent, or just a program?: A taxonomy for autonomous agents, in *Proceedings of the Third International Workshop on Agent Theories, Architectures, and Languages*, J. P Müller, M. J. Wooldridge, and N. R. Jennings, eds., Springer-Verlag, 1996, 21–36.

Fredkin, E., Digital mechanics: An informational process based on reversible universal cellular automata, *Phys. D* **45** (1990), 254–270.

Fredkin, E., A new cosmogony, in *Proceedings of the Workshop on Physics and Computation*, D. Matzke, ed., IEEE Comp. Soc. Press, New York, 1993, 116–121.

Fredkin, E. and Toffoli, T., Conservative logic, *Int. J. Theor. Phys.* **21** (1982), 219–253.

Frisch, U., Hasslacher, B., and Pomeau, Y., Lattice-gas automata for the *Navier-Stokes* equation, *Phys. Rev. Lett.* **56** (1986), 1505–1508.

Frisch, U., d'Humieres, D., Hasslacher, B., Lallemand, P., Pomeau, Y., and Rivet, J. P., Lattice gas hydrodynamics in two and three dimensions, *Complex Syst.* **1** (1987), 648–707, reprinted in *Lattice Gas Methods for Partial Differential Equations*, G. D. Doolen, U. Frisch, B. Hasslacher, S. Orszag, and S. Wolfram, eds., Addison-Wesley, 1990, 77–135.

Gale, D., Propp, J., Sutherland, S., and Troubetskoy, S., Further travels with my ant, *Math. Intelligencer* **17**, no. 3 (1995), 48–56.

Garfinkel, A., The slime mold *Dictyostelium* as a model of self-organization in social systems, in *Self-Organizing Systems: The Emergence of Order*, E. Yates, ed., Plenum Press, 1987.

Gaylord, R. and D'Andra, L., *Simulating Society: A Mathematica Toolkit for Modelling Socioeconomic Behavior*, Springer-Verlag, New York, 1998.

Gaylord, R. and Nishidate, K., *Modeling Nature — Cellular Automata Simulations with Mathematica*, Springer-Verlag, New York, 1996.

Gaylord, R. and Wellin, P. R., *Computer Simulations with Mathematica: Explorations in Complex Physical and Biological Systems*, TELOS/Springer-Verlag, 1994.

Gell-Mann, M., *The Quark and the Jaguar — Adventures in the Simple and Complex,* W. H. Freeman, New York, 1994.

Gerhardt, M. and Schuster, H., A cellular automaton describing the formation of spatially ordered structures in chemical systems, *Phys. D* **36** (1989), 209–221.

Gershenson, C., *Introduction to random Boolean networks*, `http://uk.arxiv.org/abs/nlin.AO/0408006`, August 2004.

Gilbert, N. and Troitzsch, K. G., *Simulation for the Social Scientist,* Open University Press, 1999.

Grand, S., *Creation: Life and How to Make It*, Harvard University Press, Cambridge, MA, 2001.

Grassberger, P., Critical behaviour of the Drossel-Schwabl forest fire model, *New J. Phys.* **4** (2002), 17.1–17.15.

Greenberg, J. M. and Hastings, S. P., Spatial patterns for discrete models of diffusion in excitable media, *SIAM J. Appl. Math.* **34** (1978), 515–523.

Guy, R. K., John Horton Conway, in *Mathematical People: Profiles and Interviews*, D. J. Albers and G. L. Alexanderson, eds., Birkhäuser, Boston, 1985, 43–50.

Hardy, J., de Pazzis, O., and Pomeau, Y., Time evolution of a two-dimensional model system. I. Invariant states and time correlation functions, *J. Math. Phys.* **14** (1973), 1746–1759.

Hedlund, G. A., Endomorphisms and automorphisms of the shift dynamical system, *Math. Syst. Theory* **3** (1969), 320–375.

Hegselmann, R. and Flache, A., Understanding complex social dynamics: A plea for cellular automata based modelling, *J. Artificial Soc. Social Simulation.* **1** (1998), `http://www.soc.surrey.ac.uk/JASSS/1/3/1.html`.

Held, G. A., Solina, D. H., Keane, D. T., Haag, W. J., Horn, P. M., and Grinstein, G., Experimental study of critical-mass fluctuations in an evolving sandpile, *Phys. Rev. Lett.* **65** (1990), 1120–1123.

Hénon, M., Isometric collision rules for the four-dimensional FCHC lattice gas, *Complex Systems* **1** (1987), 475–494.

Herman, G. T. and Liu, W. H., The daughter of Celia, the French flag, and the firing squad, *Simulation* **21** (1973), 33–41.

Holland, J. H., Universal spaces: A basis for studies in adaptation, in *Automata Theory*, E. R. Caianiello, ed., Academic Press, New York, 1966, 218–230.

Holland, J. H., The dynamics of searches directed by genetic algorithms, in *Evolution, Learning and Cognition*, Y. C. Lee, ed., World Scientific, 1988, 111–128.

Holland, J. H., *Adaption in Natural and Artificial Systems: An Introductory Analysis with Application to Biology, Control and Artificial Intelligence*, MIT Press, Cambridge, MA, 1992.

Holland, J. H., *Emergence: From Chaos to Order*, Oxford University Press, 2000.

Hopfield, J. J., Neural networks and physical systems with emergent collective computational abilities, *Proc. Natl. Acad. Sci. U.S.A.* **79** (1982), 2554–2558.

Huberman, B. A. and Glance, N. S., Evolutionary games and computer simulations, *Proc. Nat. Acad. Sci. U.S.A.* **90** (1993), 7716–7718.

Hulata, E., Volman, V., and Ben-Jacob, E., Self-regulated complexity in neural networks, *Natural Comput.* **4** (2005), 363–386.

id Quantique, Random numbers generation using quantum physics, white paper, 2004, www.idquantique.com/.

Ilachinski, A., *Cellular Automata: A Discrete Universe*, 2nd ed., World Scientific, Singapore, 2002.

Ingerson, T. E. and Buvel, R. L., Structure in asynchronous cellular automata. *Phys. D* **10** (1984), 59–68.

Istrail, S., Statistical mechanics, three-dimensionality and NP-completeness: I. Universality of intractability of the partition functions of the Ising model across non-planar lattices, in *Proceedings of the 32nd ACM Symposium on the Theory of Computing (STOC00)*, ACM Press, 2000, 87–96.

Johnson, S., *Emergence: The Connected Lives of Ants, Brains, Cities and Software*, Penguin, 2002.

Kadanoff, L., Coppersmith, S., and Aldana, M., Boolean dynamics with random couplings, Nonlinear Sciences archive, http://arxiv.org/abs/\\nlin/0204062,2002.

Kaplan, D. T., Smith, J. M., Saxberg, B. E. H., and Cohen, R.J., Nonlinear dynamics in cardiac conduction, *Math. Biosci.* **90** (1988), 19–48.

Kari, J., Reversibility of 2D cellular automata is undecidable, *Phys. D* **45** (1990), 379–385.

Kauffman, S. A., Metabolic stability and epigenesis in randomly constructed genetic nets, *J. Theor. Biol.* **22** (1969), 437–467

Kauffman, S. A., *The Origins of Order: Self-Organization and Selection in Evolution*, Oxford University Press, 1993.

Kauffman, S. A., *At Home in the Universe: The Search for Laws of Self-Organization and Complexity*, Oxford University Press, 1995.

Kauffman, S. A., *Investigations*, Oxford University Press, 2000.

Keller, E. F. and Segel, L., Initiation of slime mold aggregation viewed as an instability, *J. Theor. Biol.* **26** (1970), 399–415.

Kirchkamp, O., Spatial evolution of automata in the prisoners' dilemma, *J. Econ. Behav. Org.* **43** (2000), 239–262.

Knuth, D., *Seminumerical Algorithms*, Addison-Wesley, Reading, MA, 1981.

Kusch, I. and Markus, M., Mollusc shell pigmentation: Cellular automaton simulations and evidence for undecidability, *J. Theor. Biol.* **178** (1996), 333–340.

Land, M. and Belew R. K., No perfect two-state cellular automaton for density classification exists, *Phys. Rev. Lett.* **74** (1995), 5148–5150.

Langton, C., Self-reproduction in cellular automata, *Phys. D* **10** (1984), 135–144.

Langton, C., Studying artificial life with cellular automata, *Phys. D* **22** (1986), 120–149.

Langton, C., Computation at the edge of chaos: Phase transitions and emergent computation, *Phys. D* **42** (1990), 12–37.

Lewin, R., *Complexity — Life at the Edge of Chaos*, The University of Chicago Press, 1999.

Libbrecht, K. and Rasmussen, P., *The Snowflake: Winter's Secret Beauty*, Voyageur Press, 2003.

Lind, D., Applications of ergodic theory and sofic systems to cellular automata, *Phys. D* **10** (1984) 36–44.

Lindenmayer, A., Mathematical models for cellular interactions in development, *J. Theor. Biol.* **18** (1968), 280–315.

Lumer, E. D. and Nicolis, G., Synchronous versus asynchronous dynamics in spatially distributed systems, *Phys. D* **71** (1994), 440–452.

Luque, B. and Solé, R.V., Lyapunov exponents in random Boolean networks, *Phys. A* **284** (2000), 33–45.

Madore, B. F. and Freedman, W. L., Computer simulations of the Belousov-Zhabotinsky reaction, *Science* **222** (1983), 615–616.

Maeterlinck, M., *The Life of the Bee*, translated by A. Sutro, George Allen & Unwin, 1901.

Mandelbrot, B., *The Fractal Geometry of Nature*, W.H. Freeman, 1982.

Martin, O., Odlyzko, A., and Wolfram, S., Algebraic properties of cellular automata, *Comm. Math. Phys.* **93** (1984), 219–258.

Mazoyer, J., A six-state minimal time solution to the firing squad synchronization problem, *Theor. Compt. Sci.* **50** (1987), 183–238.

McCulloch, W. S. and Pitts, W., A logical calculus of the idea immanent in nervous activity, *Bull. Math. Biophys.* **5** (1943), 115–133.

McIntosh, H. V., Wolfram's class iv automata and a good life, *Phys. D* **45** (1990), 105–121.

Meinhardt, H., *The Algorithmic Beauty of Sea Shells*, Springer-Verlag, Heidelberg, 1995.

Miller, P., Swarm theory, *National Geographic* **212**, no. 1 (2007), 126–147.

Minsky, M., *Computation: Finite and Infinite Machines*, Prentice-Hall, 1967.

Minsky, M., *Society of Mind*, Simon & Schuster, New York, 1988.

Minsky, M., *The Emotion Machine*, Simon & Schuster, New York, 2006.

Minsky, M. and Papert, S., *Perceptrons: An Essay in Computational Geometry*, MIT Press, Cambridge, MA,1969.

Mitchell, M., *An Introduction to Genetic Algorithms*, MIT Press, Cambridge, MA,1996.

Mitchell, M., Crutchfield, J. P., and Das, R., Evolving cellular automata to perform computations: A review of recent work, in *Proceedings of the First International Conference on Evolutionary Computation and Its Applications* (EvCA '96), Russian Academy of Sciences, 1996, http://web.cecs.pdx.edu/˜mm/evca-review.pdf.

Mitchell, M. and Forrest, S., Genetic algorithms and artificial life, *Artificial Life* **1** (1994), 267–290.

Mitchell, M., Hraber, P. T., and Crutchfield, J. P., Revisiting the edge of chaos: Evolving cellular automata to perform calculations, *Complex Systems* **7** (1993), 89–130.

Moore, E. F., Machine models of self-reproduction, *Proc. Symp. Appl. Math. Am. Math. Soc.* **14** (1962), 17–33, reprinted in *Essays on Cellular Automata*, A. W. Burks, ed., University of Illinois, 1970.

Moore, E. F., *Sequential Machines: Selected Papers*, Addison-Wesley Longman, Essex, UK, 1964.

Moore, J. H. and Hahn, L. W., A cellular automata-based pattern recognition approach for identifying gene-gene and gene-environment interactions, *Am. J. Human Genet.* **67** (2000), 52.

Moore, J. H. and Hahn, L. W., Multilocus pattern recognition using cellular automata and parallel genetic algorithms, in *Proc. of the Genetic and Evolutionary Computation Conference* (GECCO – 2001), L. Spektor and E. Goodman *et al.*, eds., Morgan Kaufmann, 1452.

Moreira, J. and Deutsch, A., Cellular automaton models of tumor development: A critical review, *Adv. in Complex Systems* **5** (2002), 247–269.

Morita, K. and Imai, K., A simple self-reproducing cellular automaton with shape-encoding mechanism, *Artificial Life V: Proceedings of the Fifth International Workshop on the Simulation and Synthesis of Living Systems*, C. G. Langton and K. Shimohara, eds., MIT Press, Cambridge, MA, 1997, 489–496.

Moyson, F. and Manderick, B., The collective behavior of ants: An example of self-organization in massive parallelism; paper presented at the AAAI Spring Symposium on Parallel Models of Intelligence, Stanford, CA, March 1988.

Mukherjee, M., Ganguly, N., and Chaudhuri, P. P., Cellular automata based authentication, in *Cellular Automata: 5th International Conference on Cellular Automata for Research and Industry*, S. Bandini, B. Chopard, and M. Tomassini eds., Springer-Verlag, 2002, 259–269.

Myhill, J., The converse of Moore's Garden-of-Eden theorem, *Proc. Am. Math. Soc.* **14** (1963), 685–686.

Newman, W. I. and Turcotte, D. L., A simple model for the earthquake cycle combining self-organized complexity with critical point behavior, *Nonlinear Process. Geophys.* **9** (2002), 453–461.

Nicolis, G. and Prigogine, I., *Exploring Complexity: An Introduction*, W. H. Freeman, New York, 1989.

Nowak, M. A. and May, R. M., Evolutionary games and spatial chaos, *Nature* **359** (1992), 826–829.

Nowak, M. A., May, R. M., and Sigmund, K., The arithmetics of mutual help, *Sci. Am.*, June 1995, 76–81.

Packard, N. H., Lattice models for solidification and aggregation, in *Theory and Applications of Cellular Automata*, S. Wolfram, ed., World Scientific, 1986, 305–310.

Packard, N. H., Adaptation toward the edge of chaos., in *Dynamic Patterns in Complex Systems*, J. A. S. Kelso, A. J. Mandell, and M. F. Schlesinger, eds., World Scientific, 1988, 293–301.

Packard, N. H. and Wolfram, S., Two-dimensional cellular automata, *J. Stat. Phys.* **38** (1985), 901–946.

Panfilov, A. V. and Winfree, A. T., Dynamical simulations of twisted scroll rings in 3-dimensional excitable media, *Phys. D* **17** (1985), 323–330.

Penrose, R., *The Emperor's New Mind: Concerning Computers, Minds, and the Laws of Physics*, Oxford University Press, 1989.

Penrose, R., *Shadows of the Mind*, Oxford University Press, 1994.

Poundstone, W., *The Recursive Universe*, Oxford University Press, 1985.

Poundstone, W., *The Prisoner's Dilemma*, Anchor, 1993.

Prigogine, I. and Stengers, I., *Order Out of Chaos*, Bantam Books, 1984.

Propp, J., Further ant-ics, *Math. Intelligencer* **16**, no. 1 (1994), 37–42.

Pytte, E., Grinstein, G., and Traub, R. D., Cellular automaton models of the CA3 region of the hippocampus, *Network* **2** (1991), 149–167.

Reggia, J. A., Armentrout, S. L., Chou, H.-H., and Peng, Y., Simple systems that exhibit self-directed replication, *Science* **259** (1993), 1282–1287.

Reiter, C. A., A local cellular model for snow crystal growth, *Chaos, Solitons & Fractals* **23**(4) (2005), 1111–1119.

Resnick, M., *Turtles, Termites, and Traffic Jams*, MIT Press, Cambridge, MA, 1994.

Roli, A. and Zambonelli, F., Emergence of macro spatial structures in dissipative cellular automata, in *Cellular Automata: 5th International Conference on Cellular Automata for Research and Industry*, S. Bandini, B. Chopard, and M. Tomassini, eds., Springer-Verlag, 2002, 44–55.

Rosenblatt, F., The perceptron: A probabilistic model for information storage and organization in the brain, *Psych. Rev.* **65** (1958), 386–408.

Sakoda, J. M., The checkerboard model of social interaction, *J. Math. Sociol.* **1** (1971), 119–132.

Sandefur, J. T., *Discrete Dynamical Systems: Theory and Applications*, Oxford University Press, 1990.

Schelling, T. C., Dynamic models of segregation. *J. Math. Sociol.* **1** (1971), 143–186.

Schepers, H. and Markus, M., Two types of performance of an isotropic cellular automaton: Stationary (Turing) patterns and spiral waves, *Phys. A* **188** (1992), 337–343.

Schönfisch, B., Propagation of fronts in cellular automata, *Phys. D* **80** (1995), 433–450.

Schönfisch, B. and de Roos, A., Synchronous and asynchronous updating in cellular automata, *BioSystems* **51** (1999), 123–143.

Schönfisch, B. and Kinder, M., A fish migration model, in *Cellular Automata: 5th International Conference on Cellular Automata for Research and Industry*, S. Bandini, B. Chopard, and M. Tomassini, eds., Springer-Verlag, 2002, 210–219.

Schrandt, R. G., and Ulam, S. M., On recursively defined geometrical objects and patterns of growth, Los Alamos Report 3762, August 1967, also reprinted in S. M. Ulam, *Analogies Between Analogies: The Mathematical Reports of S. M. Ulam and His Los Alamos Collaborators*, University of California Press, 1990.

Schulman, L. S. and Seiden, P. E., Percolation and galaxies, *Science* **233** (1986), 425–431.

Schweitzer, F., Behera, L., and Mühlenbein, H., Evolution of cooperation in a spatial prisoner's dilemma, *Adv. Complex Syst.* **5** (2002) 269–299.

Shannon, C. E., A mathematical theory of communication, *Bell Syst. Tech. J.* **27** 1948, 379–423, 623–656.

Singh, S., *The Code Book*, Fourth Estate, 1999.

Sipper, M., Non-uniform cellular automata: Evolution in rule space and formation of complex structures, in *Artificial Life IV*, R. A. Brooks and P. Maes, eds., MIT Press, Cambridge, MA, 1994.

Sipper, M., Studying artificial life using simple, general cellular automata, in *Artificial Life* **2** (1995), 1–35.

Sipper, M., Fifty years of research on self-replication: An overview, *Artificial Life* **4** (1998), 237–257.

Smith, A. R., Simple, nontrivial self-reproducing machines, in *Artificial Life II*, C. G. Langton, C. Taylor, J. D. Farmer, and S. Rasmussen, eds., Addison-Wesley, 1991, 709–725.

Sornette, D., *Critical Phenomena in Natural Sciences*, 2nd ed., Springer-Verlag, Heidelberg, 2004.

Stauffer, A. and Sipper, M., An interactive self-replicator implemented in hardware, *Artificial Life* **8** (2002), 175–183.

Stewart, I., The ultimate in anty-particles, *Sci. Am.*, July 1994, 104–107.

Succi, S., Kinetic approach to lattice quantum mechanics, in *Cellular Automata: 5th International Conference on Cellular Automata for Research and Industry*, S. Bandini, B. Chopard, and M. Tomassini, eds., Springer-Verlag, 2002, 114–122.

Tamayo, P. and Hartman, H., Cellular automata, reaction-diffusion systems and the origin of life, in *Artificial Life I,* C. G. Langton, ed., Addison-Wesley, 1989, 105–124.

Toffoli, T., Cellular automata as an alternative to (rather than an approximation of) differential equations in modeling physics, *Phys. D* **10** (1984) 117–127.

Toffoli, T. and Margolus, N., *Cellular Automata Machines — A New Environment for Modeling*, MIT Press, Cambridge, MA, 1987.

Toffoli, T. and Margolus, N., Invertible cellular automata: A review, *Phys. D* **45** (1990), 229–253.

Tomassini, M. and Venzi, M., Artifically evolved asychronous cellular automata for the density task, in *Cellular Automata: 5th International Conference on Cellular Automata for Research and Industry*, S. Bandini, B. Chopard, and M. Tomassini, eds., Springer-Verlag, 2002, 44–55.

Turing, A., On computable numbers, with an application to the Entscheidungsproblem, *Proc. Lond. Math. Soc.* Ser. 2 **42** (1936), 230–265.

Turing, A., Computing machinery and intelligence, *Mind* **LIX**, no. 236 (1950), 433–460.

Turing, A., The chemical basis for morphogenesis, *Philos. Trans. Roy. Soc.* **237** (1952), 37–72.

Ulam, S. M., Random processes and transformations, *Proc. Int. Cong. Math.*, Cambridge, MA, 1950, Vol. 2, 264–275.

Vichniac, G., Simulating physics with cellular automata, *Phys. D* **10** (1984), 96–116.

von Neumann, J., *Theory of Self-Reproducing Automata*, edited and completed by A. W. Burks, University of Illinois Press, 1966.

von Neumann, J. and Morgenstern, O., *Theory of Games and Economic Behavior*, Princeton University Press, 1944, reprinted 1980.

Waddington, C. H. and Cowe, R. J., Computer simulations of a molluscan pigmentation pattern, *J. Theor. Biol.* **25** (1969), 219–225.

Wade, D., *Li—Dynamic Form in Nature*, Walker & Co., New York, 2003.

Waksman, A., An optimum solution to the firing squad synchronization problem, *Inform. Control* **8** (1966), 66–78.

Waldrop, M. M., *Complexity, the Emerging Science at the Edge of Order and Chaos*, Simon & Schuster, New York, 1992.

Williams, G. P., *Chaos Theory Tamed*, Joseph Henry Press, Washington, D.C., 2001.

Wolfram, S., Universality and complexity in cellular automata, *Phys. D* **10** (1984), 1–35.

Wolfram, S., Random sequence generation by cellular automata, *Adv. Appl. Math.* **7** (1986), 123–169.

Wolfram, S., *Cellular Automata and Complexity*, Westview, 1994.

Wolfram, S., *A New Kind of Science*, Wolfram Media, Champaign, IL, 2002.

Young, D. A., A local activator-inhibitor model of vertebrate skin patterns, *Math. Biosci.* **72** (1984), 51–58.

Zuse, K., *Rechnender Raum* (Vieweg. Braunschweig, 1969), translated as *Calculating Space*, Tech. Transl. AZT-70-164-GEMIT, MIT Project MAC, 1970.

Zykov, V., Mytilinaios, E., Adams, B., and Lipson, H., Self-reproducing machines, *Nature* **435** (2005), 163–164.

INDEX

R

S

T

U

V

W

X

Z

WILEY-INTERSCIENCE
SERIES IN DISCRETE MATHEMATICS AND OPTIMIZATION

AARTS AND KORST • Simulated Annealing and Boltzmann Machines: A Stochastic Approach to Combinatorial Optimization and Neural Computing
AARTS AND LENSTRA • Local Search in Combinatorial Optimization
ALON, SPENCER, AND ERDŐS • The Probabilistic Method, Second Edition
ANDERSON AND NASH • Linear Programming in Infinite-Dimensional Spaces: Theory and Application
ARLINGHAUS, ARLINGHAUS, AND HARARY • Graph Theory and Geography: An Interactive View E-Book
AZENCOTT • Simulated Annealing: Parallelization Techniques
BARTHÉLEMY AND GUÉNOCHE • Trees and Proximity Representations
BAZARRA, JARVIS, AND SHERALI • Linear Programming and Network Flows
BRUEN AND FORCINITO • Cryptography, Information Theory, and Error-Correction: A Handbook for the 21st Century
CHANDRU AND HOOKER • Optimization Methods for Logical Inference
CHONG AND ŻAK • An Introduction to Optimization, Second Edition
COFFMAN AND LUEKER • Probabilistic Analysis of Packing and Partitioning Algorithms
COOK, CUNNINGHAM, PULLEYBLANK, AND SCHRIJVER • Combinatorial Optimization
DASKIN • Network and Discrete Location: Modes, Algorithms and Applications
DINITZ AND STINSON • Contemporary Design Theory: A Collection of Surveys
DU AND KO • Theory of Computational Complexity
ERICKSON • Introduction to Combinatorics
GLOVER, KLINGHAM, AND PHILLIPS • Network Models in Optimization and Their Practical Problems
GOLSHTEIN AND TRETYAKOV • Modified Lagrangians and Monotone Maps in Optimization
GONDRAN AND MINOUX • Graphs and Algorithms *(Translated by S. Vajdā)*
GRAHAM, ROTHSCHILD, AND SPENCER • Ramsey Theory, Second Edition
GROSS AND TUCKER • Topological Graph Theory
HALL • Combinatorial Theory, Second Edition
HOOKER • Logic-Based Methods for Optimization: Combining Optimization and Constraint Satisfaction
IMRICH AND KLAVŽAR • Product Graphs: Structure and Recognition
JANSON, LUCZAK, AND RUCINSKI • Random Graphs
JENSEN AND TOFT • Graph Coloring Problems
KAPLAN • Maxima and Minima with Applications: Practical Optimization and Duality
LAWLER, LENSTRA, RINNOOY KAN, AND SHMOYS, Editors • The Traveling Salesman Problem: A Guided Tour of Combinatorial Optimization
LAYWINE AND MULLEN • Discrete Mathematics Using Latin Squares
LEVITIN • Perturbation Theory in Mathematical Programming Applications
MAHMOUD • Evolution of Random Search Trees
MAHMOUD • Sorting: A Distribution Theory
MARTELLI • Introduction to Discrete Dynamical Systems and Chaos
MARTELLO AND TOTH • Knapsack Problems: Algorithms and Computer Implementations
McALOON AND TRETKOFF • Optimization and Computational Logic
MERRIS • Combinatorics, Second Edition
MERRIS • Graph Theory
MINC • Nonnegative Matrices
MINOUX • Mathematical Programming: Theory and Algorithms *(Translated by S. Vajdā)*
MIRCHANDANI AND FRANCIS, Editors • Discrete Location Theory
NEMHAUSER AND WOLSEY • Integer and Combinatorial Optimization
NEMIROVSKY AND YUDIN • Problem Complexity and Method Efficiency in Optimization *(Translated bv E. R. Dawson)*

PACH AND AGARWAL • Combinatorial Geometry
PLESS • Introduction to the Theory of Error-Correcting Codes, Third Edition
ROOS AND VIAL • Ph. Theory and Algorithms for Linear Optimization: An Interior Point Approach
SCHEINERMAN AND ULLMAN • Fractional Graph Theory: A Rational Approach to the Theory of Graphs
SCHIFF • Cellular Automata: A Discrete View of the World
SCHRIJVER • Theory of Linear and Integer Programming
SPALL • Introduction to Stochastic Search and Optimization: Estimation, Simulation, and Control
SZPANKOWSKI • Average Case Analysis of Algorithms on Sequences
TOMESCU • Problems in Combinatorics and Graph Theory *(Translated by R. A. Melter)*
TUCKER • Applied Combinatorics, Second Edition
WOLSEY • Integer Programming
YE • Interior Point Algorithms: Theory and Analysis